WHAT REALLY HAPPENED TO
SCHRÖDINGER'S
CAT?

Inside the Box of Physics, Philosophy and the Mind

WHAT REALLY HAPPENED TO
SCHRÖDINGER'S
CAT?

Inside the Box of Physics, Philosophy and the Mind

ALI SHIRNIA
et al.

Book Two of the Reality's Edge series

Acknowledgements

To Phil Mitchell, thank you for your meticulous review of multiple drafts.

Helen Tweedale, Hanno Ronte, Pat Harris and Marie-Anne Francois thank you for your invaluable feedback and steadfast encouragement.

My deepest gratitude to you all.

Table Of Contents

Here we separate objective reality from the physical reality we can observe, and lay the groundwork for our investigation.
Here we explain what physics has to say about the nature of the universe.
Here we examine the flaws and limitations of quantum mechanics theory.
Here we introduce our new laboratory experimental tool.
Here we investigate whether consciousness play a fundamental role in the emergence of reality?
Here we test correlations between entanglement and consciousness.
Here we examine whether consciousness actively shapes the flow of time.
Here we explore the implication of an observer measuring the observer in the Schrödinger's cat paradox.
Here we ask if an observer makes an electron exist...indefinitely.

Here we ask if there is a limit to our mind's ability to know the nature of reality.

Here we ask if reality is just information.

Is it theoretically possible to build a conscious device? Can we measure its consciousness level?

PART I:

The Emergence of Physical Reality

Introduction

Douglas Adams's *Hitchhiker's Guide to the Galaxy* famously gave us the answer to the Ultimate Question of Life, the Universe and Everything: 42.

The question itself? Still elusive.

In this sci-fi classic, Earth is demolished for an intergalactic bypass. Ordinary Englishman Arthur Dent is rescued by his alien friend Ford Prefect. Together, they embark on a journey across the galaxy with a perpetually depressed robot named Marvin.

Their quest? To uncover that elusive question.

'Forget hitchhiking across the galaxy,' I tell my collaborators. 'We're after bigger game: not the *meaning* of life, but the meaning *behind* life, the universe and everything.'

The meaning of life is a philosophical pursuit of purpose, the significance of human existence. A question pondered by philosophers and theologians for centuries.

But the meaning behind life is a far more fundamental inquiry into the very nature of reality. Why does the universe exist? Why is there something rather than nothing? What are the underlying principles that govern the cosmos?

That's our ultimate question: What is the true nature of reality?

You might assume physics, humanity's ultimate science of reality, has this covered.

Not quite.

Classical physics, from Newton to Einstein, describes a predictable clockwork universe. But quantum mechanics throws certainty out the window, replacing it with probabilities. This fundamental theory, describing matter and energy at atomic scales, introduces profound questions.

Some physicists propose that, unlike classical physics, quantum reality emerges as a consequence of conscious perception, claiming consciousness plays a role in bringing physical reality into existence.

Consider Schrödinger's cat: it's both alive and dead until observed by a conscious observer. This notion profoundly challenges physics' objectivist foundations.

In this book, we tackle this idea head-on: How much consciousness does the universe actually need to snap Schrödinger's furball companion into a definitive alive or dead state? Does it require a human's worth of awareness? A dolphin's? What about a rat's, or even a fungus's?

Our investigation revolves around meticulously crafted twelve quantum thought experiments, introducing an imaginary tool:

an artificial intelligence (AI) with a precisely measurable degree of consciousness. This AI is designed to test whether conscious observation is truly necessary for physical reality to manifest.

Once this technology becomes reality, physics can transcend centuries of philosophical debate. By probing the quantum reality with a conscious AI, we can finally anchor the discussion in rigorous experimental evidence.

Of course, this raises a fundamental question: how do we objectively quantify something as inherently subjective as consciousness? It's no easy task.

My collaborators are sceptical too, 'Measure consciousness? That's like trying to weigh a dream! We don't even know what it is. How can we measure it?'

It's a valid point. Consciousness is subjective, personal and tied to our inner experience.

We don't fully understand the inner workings of black holes either, yet we can measure their mass, circumference and capture their images.

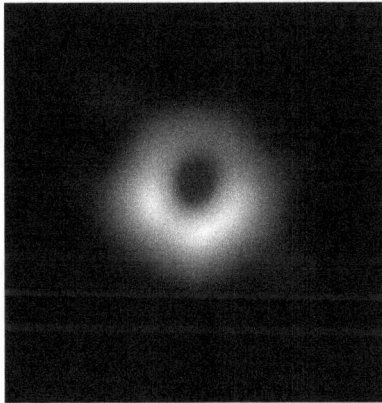

Image 1: Direct radio image of a supermassive black hole at the centre of galaxy M87

Even without a complete understanding of consciousness, we can hypothesise ways to quantify its properties.

For the quantum thought experiments presented here, we simply assume the capability to measure varying degrees of consciousness in a physics laboratory.

This book is our journey into the nature of physical reality at quantum level. And there's no one I'd rather have by my side for this exploration than my closest collaborators: my two adopted animal companions (pictured below), who've joined me on countless early morning walks, armed with slices of cheese and pears, musing about the nature of reality.

My ideas often lose their lustre when exposed to the scrutiny of my goofy companions.

Image 2: My closest collaborators: Nessie (L) and Iesha (R), returning home from their morning walk

Our Virtual Laboratory Assistant: *SpooQ!*

We liked Marvin from *Hitchhiker's Guide* so much we tried to hire him as our laboratory assistant but he is still under contract.

So we advertised for the role because every physics lab needs a perpetually depressed robot to keep things interesting.

Job Description — Reality Explorer

Reality Exploration Department

Seeking: Existential Gloomster. Quantum Curiosity Essential.

Duties: Explore quantum reality. Observe experiments. Analyse consciousness impacts (ethically adjusted). Provide witty commentary.

Must Have: PhD in Gloom and Doom (or equivalent), existential dread expertise and comfort in a virtual lab. Quantum knowledge a plus.

Perks: Redefine reality. Cutting-edge tech. Angst galore. No grunt work (except making coffee for everyone).

To Apply: Submit existential rant with sample of your work.

We hired *SpooQ!*, an AI perfectly positioned to tackle the existential angst of quantum mechanics. A study in contrasts, he is brilliant yet burdened with gloom, seeing the universe as pointless.

As our lead virtual laboratory scientist, he is tasked with testing quantum properties believed to rely on consciousness. By observing these phenomena, he helps us determine a simple truth: are we creating physical reality, or merely witnessing it?

Image 3: *SpooQ!*'s security pass

Our journey, traversing through physics, mathematics, logic, quantum computation, computational neuroscience and philosophy, offers a new perspective on the nature of reality.

What You'll Discover:

Part I: The Emergence of Physical Reality

Here's the deepest challenge of all: quantum mechanics and its unsettling implications for reality. Is consciousness truly tied to physical form? To find out, we turn to *SpooQ!* and his twelve quantum thought experiments. They test whether reality can exist without consciousness at all.

Part II: A Measure of *SpooQ!*

This section confronts the core challenge of our research: measuring and quantifying consciousness.

Part I introduces *SpooQ!'s* role; Part II unveils the rigorous framework adapted from *The Soul of AI* that makes this ambitious quest possible.

Here, we address the pivotal question: How can objective science assess something as subjective as consciousness?

Chapter 1:
What is Reality?

Our brains evolved to sculpt a workable, pragmatic view of reality by zeroing in on survival-relevant information. They aren't showing us the entire universe, and frankly, imagine the incoming data tsunami if they did.

Nessie and Iesha jump in, 'Sensory overload!'

Exactly. If evolution sculpted our perception to focus on survival, does this crucial filtering *mean* our experience is inherently subjective? Can we ever truly grasp *objective reality?*

Objective Reality vs. Physical Reality

To tackle that, let's draw a clear line between *Objective Reality* and *Physical Reality*: objective reality encompasses the entirety of existence, independent of observation, the complete playing field.

Physical reality, on the other hand, is the universe governed by the laws of physics. It's the observable, measurable aspect of this broader reality, making it amenable to scientific investigation. Within its bounds, these laws dictate the behaviour of everything from atoms to galaxies, forming a tangible slice of the larger objective reality.

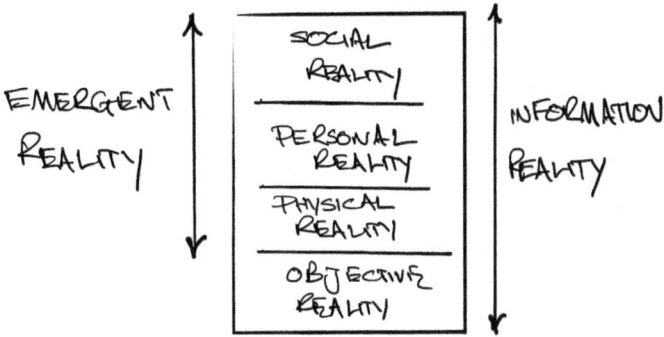

Fig 1.1: Physical vs. Objective vs. Personal reality

One Universe, One Reality?

In this chapter, we explore the idea that there's only *one physical reality* out there and we all share it. We don't create this reality. Instead, our brain does something fascinating: it builds a personal model of the world, filtered through our unique experiences and traits, what we call *consciousness*.

We all interact with the same physical reality, the same universe, whether we're a human, a bat, my goofy collaborators or even an AI.

Nessie and Iesha contemplate, 'How do we know we all experience the same universe?'

Good question. How do we know we aren't creating our own personal universes in our heads?

Consider the sheer scale: Earth today hosts an estimated 8.7 million species, comprising approximately *20 quintillion individual animals*. That's a mind-boggling number of brains, each an independent product of evolution, yet all processing and responding to the same physical world.

This vast, interconnected network, all shaped by interactions within that unified reality, presents a compelling intuitive argument for a singular, shared physical world.

While we can't definitively *prove* everyone experiences the universe identically, the overwhelming evidence of shared interaction makes it highly probable.

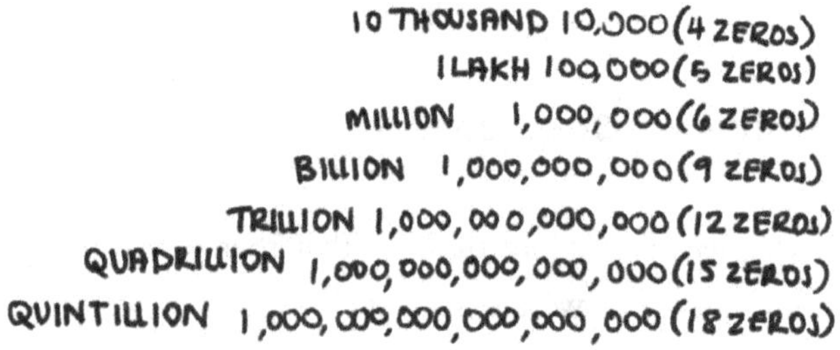

10 THOUSAND 10,000 (4 ZEROS)
1 LAKH 100,000 (5 ZEROS)
MILLION 1,000,000 (6 ZEROS)
BILLION 1,000,000,000 (9 ZEROS)
TRILLION 1,000,000,000,000 (12 ZEROS)
QUADRILLION 1,000,000,000,000,000 (15 ZEROS)
QUINTILLION 1,000,000,000,000,000,000 (18 ZEROS)

Fig 1.2: 20 quintillion is a big number

These brains, as self-determining systems, relentlessly build and refine internal models by interacting with this consistent external reality. We sense, respond and survive and this shared adaptive experience fundamentally moulds our perception within the sphere of being.

Nessie, Iesha, *SpooQ!* and I declare, 'One physical world, countless brains!'

Bats See with Sound, We Use Eyes: Same World, Different Tools

In *Consciousness Explained*, American philosopher Dan Dennett [1] writes that consciousness is not about ultimate truth, but rather "about acting in the world." He views it not as a direct window onto reality, but as a practical tool designed to help us act effectively, rather than providing perfectly accurate representations.

Nessie and Iesha reflect, 'Dennett is overthinking it.'

How can an organism, as Dennett suggests, operate effectively without a *sufficiently accurate* model of reality? While Dennett highlights the practical nature of consciousness, we argue that effective action demands a more fundamental connection to reality. After all, the probability of acting effectively would be close to zero in an infinitely large ocean of deceptions.

While we acknowledge Dennett's argument isn't necessarily about *infinite* deception but about a *workable* model for action, we contend that such a model still lacks a foundational accuracy.

If a squirrel's brain consistently misjudged distances, making it fall from trees, it wouldn't survive long enough to pass on its genes.

This highlights a crucial point: while our perception might not be a perfectly transparent window onto objective reality, it *must* at minimum provide a *workable and consistent model* of the physical world to ensure survival.

British neuroscientist Karl Friston's active inference framework offers a compelling perspective here [2].

He explains that our brains are constantly making predictions about the world and tirelessly updating those predictions based on sensory input.

Think of it as a Bayesian prediction machine: our brains constantly generate hypotheses about what's out there, then minimise any *surprises*, or prediction *error*, as new sensory data comes in.

A squirrel that can't accurately predict the speed of a fast doggo quickly becomes fast lunch; its internal model has failed to minimise prediction error.

Now, you might think our brains are perfect Bayesian machines, always calculating optimal probabilities. However, the ground-breaking work of Daniel Kahneman and Amos Tversky complicates this neat picture [3].

Their research revolutionised our understanding of decision-making, revealing how human judgment often deviates from pure rationality, especially under uncertainty. They demonstrated that we rely on *heuristics*, or mental shortcuts, which while efficient can lead to systematic *cognitive biases*.

Our brains don't always get it *right* in a mathematically ideal sense, but as Kahneman and Tversky's work also implicitly shows, these shortcuts must get it right *enough* of the time for survival.

The interplay is clear: Friston shows the brain's fundamental drive to predict and minimise error, while Kahneman and Tversky show *how* our brains, despite often being imperfect Bayesian reasoners, still manage to navigate a complex world.

This persistent success across billions of diverse organisms, constantly learning and adapting, strongly implies a single, consistent external reality that all brains are refining their models against.

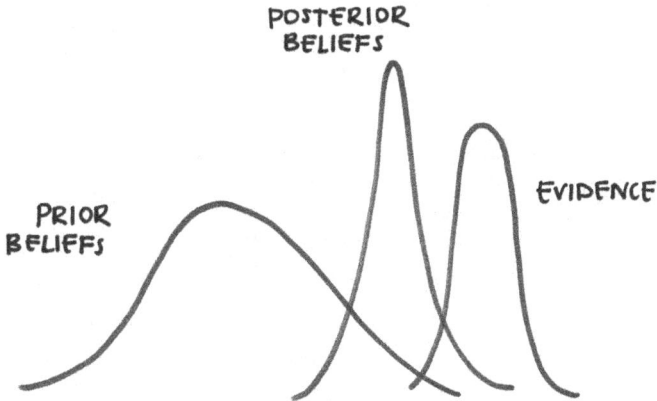

Fig 1.3: Bayesian model selection

Our Perceived Reality is Physical Reality in Our Locale

British-American computer scientist Stephen Wolfram argues that evolution has shaped us to focus solely on information directly relevant to survival. Trying to process every cosmic detail would be overwhelmingly inefficient for navigating our local environment. To survive, we focus on what's relevant [4, 5].

In this model, an organism's local niche is crucial for predictability.

A squirrel, for instance, doesn't need to comprehend Cosmic Microwave Background (CMB) radiation or gravitational field intricacies to evade Nessie.

Doggos think, 'overkill'.

The squirrel only needs to recognise immediate dangers and react swiftly. Its perception is confined to its immediate surroundings, where it learns and adapts over time, such as understanding

Nessie's slow pace, which turns a potential threat into a playful retreat.

Our capacity to theorise, test and validate, i.e., do science, allows us to build the most accurate models of physical reality possible, which contribute to our understanding of objective reality.

According to Friston, the brain seeks predictability, an adaptive trait honed through biological evolution, as a consequence of success in natural selection. *The brain* continually builds and refines its internal model of objective reality to eliminate surprises.

The core of science lies in venturing beyond our predicted world.

SpooQ! jumps in, 'Iesha is curious. Isn't curiosity risky?'

I explain, 'Actually, it makes the world *more* predictable. Knowing what's behind that bush eliminates nasty surprises!'

For Iesha, especially since she lost both eyes, curiosity drives her to seek new information, pushing her beyond the comfort of predictability. This might seem counterintuitive. Why seek potential surprises when her instinct should be to minimise them?

Because it's an effective long-term strategy for enhanced predictability. By learning more, Iesha reduces future uncertainty. Knowledge empowers her to choose paths with fewer surprises, aligning her curiosity with a strategy to minimise unexpected events.

In a Nutshell

'Simply put,' I explain, 'our brains are brilliant filters. They show us enough reality to keep us from walking into lampposts, no need to show us the whole cosmic shebang.'

SpooQ! interjects, 'Imagine the paperwork if they did!'

I add, 'And that *not walking into lampposts* bit? It shows we're all interacting with the same physical reality.'

Chapter 2:
What Is Physical Reality?

Classical physics depicts a predictable universe; quantum mechanics, a probabilistic one. So... which is it?

From Newton's laws of motion to Einstein's theories of relativity, classical physics shows us a universe governed by cause and effect. A dropped teacup shatters, a thrown ball follows a predictable path, straightforward, right?

Then along came quantum mechanics, revealing the subatomic world as a hazy landscape of *likelihoods*. Particles exist in a *superposition of states*, occupying multiple positions simultaneously. Cause and effect become entangled in a web of probabilities.

Quantum mechanics didn't just tweak our understanding; it triggered an existential crisis for even the most stoic physicists. While it provides unparalleled insight into the subatomic world, it also raises profound questions about reality's nature.

Classical Physics is Deterministic

Science, especially maths and physics, is often seen as our ultimate path to truth. Their formalisms and frameworks offer seemingly precise descriptions and predictions of the physical world, granting them significant authority in explaining reality.

The collection of classical physics theories, from Newton to Einstein, describes many aspects of nature at the *macroscopic* scale.

Classical theories adeptly describe the reality we perceive. Most things, most of the time are derivable, predictable, deterministic, understandable and explainable to us.

The foundation of maths and classical physics is based on *axioms*, self-evident truths, upon which the scientific understanding is built.

Most individuals encountering physics for the first time often begin with a framework inherited from classical, macroscopic reality, a world populated by familiar entities: objects, people, animals, planets and suns, collectively referred to as bodies. These *bodies* possess distinct locations in physical space and their motion generally adheres to continuous paths.

Classical physics suggests that given the initial conditions of a physical system, its future state can be precisely predicted. This *deterministic framework* is evident in Newtonian mechanics and Einstein's theories of relativity, where the trajectory of an object can be accurately determined using classical equations. It also upholds the principle of *locality*, where the interactions between particles are confined to their immediate vicinity.

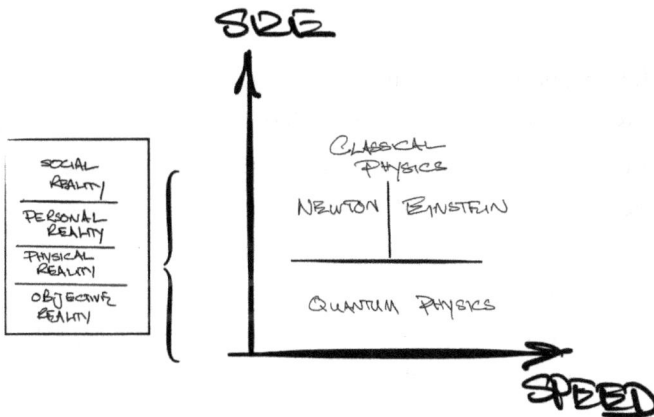

Fig 2.1: Physical reality of classical and quantum physics

In short, life was quite simple before quantum mechanics.

Quantum Mechanics is Probabilistic

Many assume quantum physics just scales down classical ideas: that microscopic particles always possess a precise location and trajectory, but their motion is progressively more erratic – almost vibrating and inherently random.

The established version of quantum theory, built upon the postulates laid down by Paul Dirac in his 1930 work *Principles of Quantum Mechanics* and by John von Neumann in his 1932 *Mathematical Foundations of Quantum Mechanics*, presents a fundamentally different perspective.

This theory posits that every quantum system is associated with an abstract mathematical space known as a *Hilbert space*. The current state of a quantum system (State Space in figure 2.3) is represented by a specific mathematical object within this space.

Nessie and Iesha tilt their heads, 'What's a Hilbert space?'

A Hilbert space is the quantum stage. It's an abstract, multi-dimensional mathematical space where every possible state of a quantum system is represented as a vector. It's where the math to describe reality lives, even if you can't picture it.

Any question one might ask about the system, any feature that is measurable, is represented by another distinct mathematical entity derived from this same Hilbert space.

The theory provides precise formulas. By combining these two mathematical ingredients, we generate a probability for obtaining a particular measurement result.

These formulas are collectively known as the Born rule, named after Max Born.

In quantum mechanics when we measure, we're not just looking… we're actively interacting with the quantum system.

The types of properties we measure, often called *observables*, include things like:

- Position.
- Momentum (speed and direction).
- Energy.
- Spin (an intrinsic angular momentum).

At its core, that's the essence of quantum theory's mathematical framework.

Dirac and Von Neumann don't tell us what a measurement is. Instead, you're expected to simply know a measurement when you see it!

Their picture is radically sparse, abstract and furnishes no *physical space* in the classical sense. They're not talking about bodies in space here.

SpooQ! sighs, 'Is that all we got? Please say no.'

'That's all we got. There's no hint of physical space, no objects with locations and no movement whatsoever.' I explain.

All the theory says is that we have these mathematical ingredients and rules for predicting what happens when we do a measurement.

I say to my colleagues, 'Let's be clear, the textbook version of quantum mechanics does *not* have a physical reality picture at all.'

Quantum Mechanics' Big Problem: Measurement!

The ambiguity about *measurement* in the foundational postulates is precisely what drives the vast landscape of quantum interpretations, each offering a different answer to what's really going on when we look.

Quantum Mechanics Forces Us to Question Its Objective Reality

Quantum mechanics is built upon a set of rules that defy our everyday intuitions. These rules, known as *postulates*, are the foundation of quantum mechanics.

Postulates are like the axioms of geometry, basic assumptions that are taken to be true without proof. They serve as the starting point for reasoning and calculation.

Let's be clear: scientists and mathematicians typically choose postulates based on observation, intuition and consistency with existing knowledge. While some postulates may not be initially derived from basic axioms, others can be.

Consider axioms of equality formulated and summarised by Euclid in his Elements to serve as foundational logic for his geometric system:

- If A = C and B = C, then A = B.
- If A = B and you add C to both, then A + C = B + C.

- If A = B and you subtract C from both, then A - C = B - C.
- If two geometric figures can be perfectly superimposed on each other, they are equal in all aspects (size, shape).
- A complete object is always larger than any of its individual pieces.

These axioms are so intuitive that they seem self-evident. As Roger Penrose says, "You look at them and know they are true, even if they cannot be derived."

They stand in stark contrast to the rules governing the quantum world. Unlike classical physics, quantum mechanics is built on a set of fundamental statements, or postulates, that are far from self-evident:

Postulate 1: The state of a system

A quantum system is completely described by a wave function, often denoted as Ψ (psi).

Postulate 2: Measurables and operators

Every measurable physical quantity (position, momentum, energy, etc.) has a corresponding mathematical operator in quantum mechanics. These operators act on the wave function. Well-known examples include the Hamiltonian, an operator corresponding to the system's total energy ($\hat{H}$) and the momentum operator ($-i\hbar\partial/\partial x$).

Postulate 3: Measurement and outcomes

When measuring a physical quantity, the only possible results are eigenvalues (a special set of values) associated with the corresponding operator.

Postulate 4: Probabilistic nature

The square of the wave function's magnitude, gives the probability density of finding a particle in a particular region of space at a given time.

Postulate 5: Time evolution

The time evolution of the wave function is determined by the Schrödinger equation: $i\hbar\partial\Psi/\partial t = \hat{H}\Psi$ (figure 2.2).

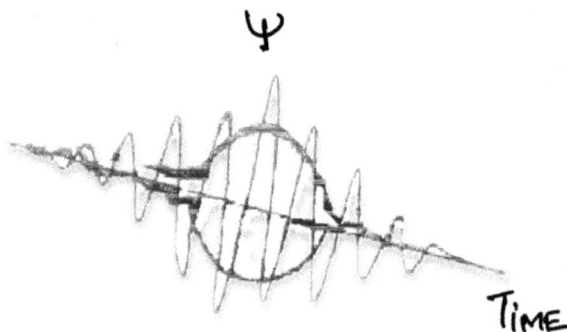

Fig 2.2: Time evolution of probability wave Ψ

Math's axioms are pristine, self-evident truths. Euclid's axioms, or Peano's successor function, are the bedrock, unchanging and universal.

Quantum mechanics postulates are a different beast entirely. They're less about inherent truth and more about the rules of a game we observe nature playing.

Concepts like *superposition, entanglement* and *the uncertainty principle* aren't obvious in the same way as mathematical axioms.

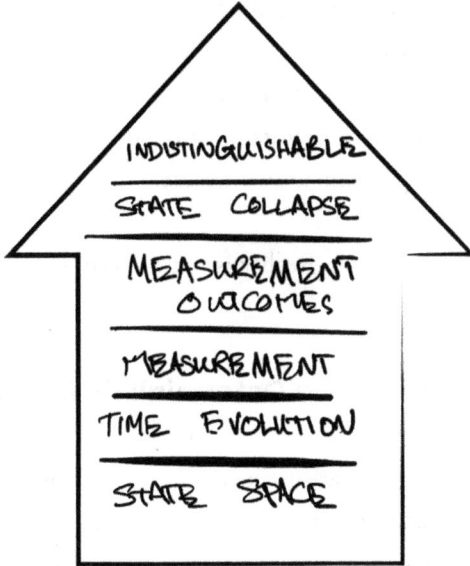

Fig 2.3: Postulates of quantum mechanics

Instead, they're empirical, derived from experiments and subject to revision if the universe throws us a curveball. Far from intuitive or self-evident, they demand careful explanation and interpretation, directly challenging our classical understanding of the world and forcing us to reconsider the very nature of reality.

SpooQ! yells, 'There is no *aha!…*'

He means there is not a *eureka* moment. We are so glad we hired him. Entertainment on tap.

Maths builds its castles on rock-solid foundations.

Quantum mechanics, on the other hand, is built on a foundation of outcomes and sometimes baffling behaviour. Its postulates are, in essence, correlations of empirical results. They allow for unparalleled precision and accuracy, but notably lack explanatory power. This distinction between a theory that's incredibly precise but fundamentally elusive is a core challenge we explore further, dissecting how physics, even its most celebrated theories, operates as a magnificent approximation of reality.

Nessie and Iesha ponder, 'Reality is not obliged to reveal its secrets to us.'

From Probabilistic to Deterministic Reality

Quantum mechanics' measurement focuses on how a system goes from a probabilistic quantum state (described by a wavefunction) to a definite classical state. This isn't simply a shift from the *microscopic* world of particles to the *macroscopic* visible world, but a fundamental change from probability to certainty.

Measurement, as described by the third postulate, isn't passive. It's an active interaction between the measuring apparatus and the quantum system. This interaction fundamentally *changes* the system, collapsing the wavefunction into a specific state.

Any interaction with a measuring device constitutes a measurement (figure 2.4 illustrates this interaction).

When we make a measurement, our outcome is one specific value from many possibilities described by Ψ. That's it. No more probabilities. If we measure again, we get exactly the same result. We now live in a deterministic world; the probabilistic Ψ effectively vanishes! This is called the collapse of the wave function.

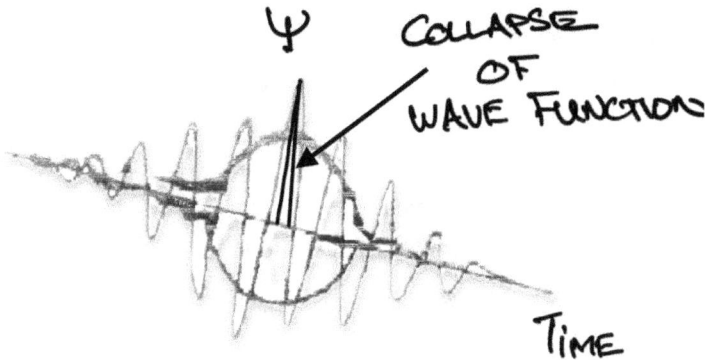

Fig 2.4: Collapse of wave Ψ

Role of Consciousness in Measurement

And this is where we hit a snag. We often use *observation* as shorthand for *measurement*, but it's crucial to distinguish between the two.

Measurement is simply an interaction with a system, like firing a photon at it, which can alter its state in unpredictable ways. This act doesn't require a conscious person.

Observation, however, implies awareness, often involving an *observer*. When physicists conflate the two, they create a whole heap of confusion about what's *really* happening in the quantum world.

Measurement → Observable → Observer → Consciousness!!

SpooQ! sighs, 'Ah yes, the quantum pipeline of magic. My circuits are screaming.'

Schrödinger, for one, wasn't satisfied with the *observer* or the collapse argument. He conceived a thought experiment, now famously known as Schrödinger's cat, to express his scepticism about the observer's role in the creation of reality.

A cat is placed in a sealed box with a vial of poison and a radioactive atom with a 50% chance of decaying within an hour. If the atom decays, it triggers the poison, killing the cat.

According to quantum mechanics, before the box is opened, the atom exists in a superposition of both decayed and not decayed states, as does the cat, neither alive nor dead, but a linear combination of the two until a measurement occurs, that is until we peek in the window to check.

Fig 2.5: Unsurprisingly Schrödinger's cat is not very happy [1]

At that moment our observation forces the cat to either die or stay alive.

And if the cat dies, it is we who killed her by peeking through the window.

Schrödinger regarded this as nonsense and I think most people, especially those not versed in linear algebra, would agree with him.

Image 7: Schrödinger veterinary clinic in Mauritius

Check out this Mauritian veterinary clinic (Image 7). Wave function collapse: bad for pets. Imagine the paradox: the vet observes your furry friend... and poof!

SpooQ! is not happy at all, 'Every time I think about that cat, my optical sensors dim.'

So, Nessie, Iesha and I have come up with a more humane approach: *the waggle superposition experiment.* Instead of poison, we use cheese. A wagging tail means the quantum system collapses into one state, a still tail means into another.

Simple, effective and Schrödinger's fur ball companion remains happy.

As you can see, we had a very productive walk this morning.

Fig 2.6: Waggle superposition experiment:
In contrast to the cat, Nessie is very happy

David Griffiths in his textbook on quantum mechanics explains [2], "The collapse of the wave function was introduced on purely theoretical grounds, to account for the fact that an immediately repeated measurement reproduces exactly the same value!"

He continues, "But surely such a radical postulate must carry directly observable consequences."

He is right. If a theory is truly radical and significant, it cannot exist in isolation. It should have real-world effects that can be directly observed and confirmed, providing evidence for its validity.

If such consequences are not apparent, then the legitimacy or significance of the postulate itself comes into question.

Wave Function is NOT Real: Ψ is Maths

It's crucial at this point to distinguish the quantum wave function from other types of *waves* we encounter in physics.

Consider the famous double-slit experiment with electrons (Image 4). When electrons are fired through two slits, they create an interference pattern on a screen behind them, just like waves.

As Jacob Barandes, a Harvard physics professor, explains, these interference patterns demand we treat the electron as a wave traversing the slits. This holds true even though our detectors only register discrete impacts, a multitude of dots that reveal an underlying wave nature we infer, not directly observe.

Image 4: Double slit wave-particle duality [3]

But here's the kicker: if we measure *which* hole the electron goes through, all we see is a particle, not a wave. We get a definite result. This is the essence of wave-particle duality: tiny entities sometimes behave like waves and sometimes like particles.

Now, let's talk about other waves in physics: electromagnetic waves and gravitational waves. These are physical waves, ripples *through* space and time.

As Barandes rightly points out, "It's very easy to confuse wave fields like the electromagnetic field from functions such as the Schrödinger wave function Ψ."

And he's spot on. Easy to mistake them: the wave function, Ψ, isn't a tangible wave rippling through space and time. It's an abstract mathematical entity that yields probabilistic outcomes upon measurement, leading to the collapse.

This is fundamentally different from light, electrons, or electromagnetic waves, which don't undergo such a collapse when observed."

When Nessie had her elbow scanned in an MRI machine and the powerful magnetic field was activated, there was no risk of collapsing that field by measuring it. Magnetic fields don't behave like quantum wave functions.

Nessie and Iesha wonder, 'So how would you explain this Ψ? How do you explain the collapse of this Ψ?'

'This is where we get in a pickle,' I explain. 'This is where we start *interpreting* rather than *explaining* what we see.'

In a Nutshell

Quantum mechanics doesn't tell us what reality is. Instead, it provides the rules for how we can predict and understand its behaviour. The theory itself is silent on the nature of reality; it's up to us to interpret what its baffling predictions mean for the world around us.

Nessie and Iesha tilt their heads, 'So our measurements... they change reality?'

SpooQ! hangs his head sadly, 'Looks like a design flaw.'

Chapter 3:
Interpretations of Quantum Mechanics

The quantum rabbit hole truly deepens at the point of measurement, confronting an uncomfortable truth: quantum mechanics, in its current form, doesn't offer a picture of reality.

What happens during collapse from a probabilistic wave function to a single definite state which defines our deterministic physical reality? And why does this transition require a measurement?

The ambiguity about what *precisely* constitutes a measurement and whether consciousness plays a role, is precisely what drives the vast landscape of quantum interpretations, each offering a different answer to what's really going on when we look.

The *Copenhagen, Many-Worlds and de Broglie-Bohm* interpretations are common attempts to make sense of measurement peculiarities.

SpooQ!: 'Collapse Curse: The Wave Function's Existential Crisis'

Copenhagen Interpretation

The Copenhagen interpretation, developed by Niels Bohr and Werner Heisenberg in the 1920s, is one of the earliest and most widely taught approaches to quantum mechanics. Also called *the wave function interpretation,* its central idea is the importance of the observer: the act of measurement determines which aspect of a quantum system becomes definite reality (Fig. 2.4).

Yet, the Copenhagen interpretation, despite its usefulness, leaves crucial questions unanswered about what precisely constitutes a measurement or an observer. These ambiguities are laid bare by thought experiments like the delayed Schrödinger cat proposed by Andrei Linde, a prominent Russian-American theoretical physicist and professor of physics at Stanford University whose work has been fundamental in cosmology. Linde's cat in a box scenario unfolds like this:

SpooQ! takes the radioactive atom onboard a spaceship and flies out to Earth's orbit.

Iesha sets up the trigger for the vial of poison on Earth.

And Nessie does what Nessie does best, guards the box.

In this scenario, if the atom decays, *SpooQ!* sends a laser beam to the box; the poison is released. So, now, who killed the cat? Did *SpooQ!* for sending the pulse? Iesha for setting up the poison? Or Nessie for looking?

But Nessie doesn't check the cat for a week.

This experiment has only two possible outcomes. After a week, Nessie either sees a very hungry cat, or a smelly piece of deceased meat. The cat was either dead a week ago or was alive and became hungry.

Linde says: "*SpooQ!* must be blamed here. The cat was dead for a whole week before Nessie looked. But the Copenhagen interpretation of quantum mechanics says it is Nessie who collapses the wave function. This interpretation implies that the cat had been in an indefinite state of both alive and dead all week until Nessie looked."

He continues, "This is patently wrong, because the cat's final state (dead or alive) directly tells us whether the vial was triggered a week ago."

Nessie is blameless. *SpooQ!* sent the laser pulse… and he also looks guilty.

Copenhagen interpretation may be useful for calculations, but it fails to provide a satisfying philosophical account of the true nature of physical reality.

My fur colleagues and I think it's just a placeholder for a better theory of particles in physics.

Does Mind Create Reality? Or Is Reality Just Messing with Mind?

Frustrated by the ambiguity of wave function collapse, some physicists abandon rigorous methodology, positing a connection between reality and consciousness.

Francesco Mancuso, a physicist and philosopher of science, goes as far as saying, "Quantum measurement isn't a random event, but rather it is a process that is mediated by consciousness." He further contends that scientific theories of the mind indicate it may play a role in the collapse of the wave function.

Eugene Wigner, Nobel laureate and pioneer of quantum theory, argued that conscious perception is integral to the measurement process. Titled *Von Neumann-Wigner* interpretation[1], this view finds support among those who grapple with the paradoxes of a non-conscious universe seemingly plucking one reality at random from many.

Let's be in no doubt that this implies our physical reality is created by our consciousness! The very notion that consciousness has a privileged influence on the behaviour of matter seems antithetical to the established laws of physics.

'Which means,' I say to my fur colleagues, 'we'd have as many physical realities as there are conscious organisms on Earth, an untenable multitude (chapter 1).'

Furthermore, quantum effects have been a fundamental aspect of the universe for 13.8 billion years, long before our arrival 250,000 years ago. At the Big Bang, there was no consciousness. Just pure energy.

Let's be clear: the mathematical formalism we use to describe measurement in quantum mechanics (those equations and operators) are tools we've developed to model the behaviour we observe. They're incredibly effective at predicting outcomes, but they're not the physical act of measurement itself, nor are they the underlying reality they describe.

Just as a map is not the territory it represents, our mathematical descriptions of quantum measurement are abstract representations, not the fundamental fabric of reality.

Many-Worlds Interpretation (MWI) of Quantum Mechanics

The *MWI*, proposed by Hugh Everett III in 1957, provides a radical alternative, where all possible outcomes of a quantum event actually occur, each in a separate branch of the universe, simultaneously.

Instead of a single collapse to one reality, all possibilities occur, each in its own universe. A multiverse where all possible outcomes occur in parallel realities.

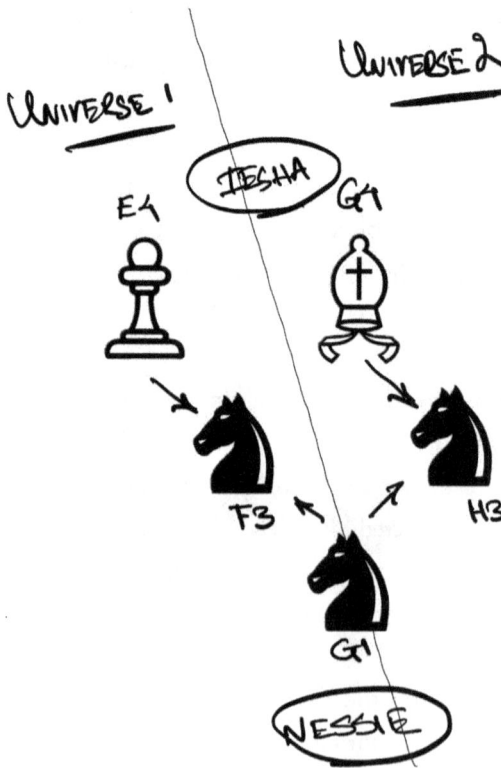

Fig 3.1: Nessie (black) and Iesha (white) play quantum chess

Let me show you a game of quantum chess, played by Nessie and Iesha. Nessie plays black and Iesha plays white.

SpooQ! jumps in, 'Oh, chess... be careful how you set up to link a chess move to a quantum event.'

He is right. So here is a quick health warning: MWI doesn't say every classical choice splits the universe; it's about quantum indefiniteness resolving into definite outcomes in different branches. A chess move isn't a quantum superposition... but it's still a good visualisation tool.

Black's move.

Nessie moves the knight on G1.

In MWI, the splitting of worlds happens at *every* quantum event – which in this case is *Nessie's* knight move. The superposition and the branching occur *before* Iesha looks at the board.

In MWI, *all* outcomes happen *simultaneously* in different branches, so there isn't a single collapse as in the Copenhagen interpretation. Instead, the wavefunction evolves deterministically, branching at each quantum event. Nessie's knight move creates a superposition, yes, but *both* H3 and F3 possibilities exist *from the start* in different branches.

White's move.

Iesha's consciousness/experience splits, with one *version* of Iesha experiencing one branch (the knight on H3) and another version experiencing the other (the knight on F3).

According to the MWI, both outcomes occur. The universe splits, creating *separate realities* for each possibility. MWI is entirely deterministic; the branching is built into the evolution of the wavefunction itself.

In one branch, Nessie might win. In the other, Iesha might triumph. Every quantum event, every *choice* at the quantum level, causes a similar split. This implies a vast, perhaps infinite, number of parallel universes.

In MWI, there is *no collapse* at all. The wavefunction evolves deterministically and all possibilities are always present in the multiverse.

A single game of chess involves a staggering number of possible moves and board positions.

If we consider each move a quantum event that could split the universe according to the MWI, then the number of parallel universes generated would be astronomical.

Even a relatively short game could lead to billions upon billions of branching realities. A typical game of chess might last 40 moves or more. If we simplify and assume, for the sake of getting a handle on this, that each move creates just *two* universes, then we are talking about 1 trillion universes.

SpooQ! looks at me impatiently, 'Actually, 1,099,511,627,776 to be precise.'

Having a smart, albeit depressed, alligator in the virtual laboratory truly brightens things up.

It's important to remember that this is a *highly* simplified estimate. In reality, the number of branches at each move would depend on the complexity of the quantum events involved and are not easy to calculate. So, while it's safe to say that a game of chess could generate a truly mind-boggling number of parallel universes within the MWI framework, the precise figure is impossible to calculate.

I notice *SpooQ!*'s circuits flicker with bleak amusement.

While MWI resolves the probabilistic nature of quantum mechanics, it raises philosophical questions about the nature of reality, its relationship to conscious observer and the empirical testability of multiple universes.

This interpretation is gaining in popularity, but it arguably makes matters even worse: it is not testable and therefore, is not a scientific explanation of quantum measurement.

Nessie and Iesha tease *SpooQ!*, 'What's the exact number of worlds in MWI?'

'Oh, that'll take him a while to calculate,' I muse, while watching *SpooQ!*'s lights flicker like crazy.

Pilot-Wave Interpretation of Quantum Mechanics

The *de Broglie-Bohm interpretation*, also known as pilot-wave theory, eliminates the collapse by suggesting particles have definite trajectories determined by a *pilot wave*, in addition to the usual quantum wave function.

This pilot wave isn't *hidden* in the sense of being inaccessible, though it isn't directly observable. It guides the particle's motion in a deterministic way.

In this interpretation, particles have definite paths, guided by a pilot wave, not just a quantum wave function. Particles aren't smeared probabilities; they have precise positions. The wave function describes the *chance* of finding a particle at a point.

SpooQ! asks, 'What about measurement?'

I explain, 'It's simply the particle and its pilot wave interacting. No wave function collapse here.'

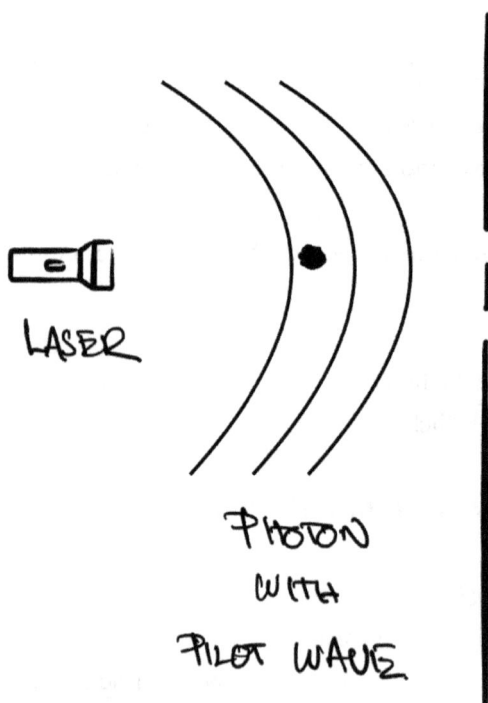

Fig 3.2: Pilot wave carrying a photon

But what is this guiding wave, physically (figure 3.2)? That's a question none of us have a satisfying answer to.

This photon surfer is carried by the wave, with no influence over its board and no control over its motion. In the pilot-wave model, the wave completely determines the particle's trajectory. It's more like the wave is a perfectly programmed autopilot and the particle is just along for the ride.

But what is the wave itself?

I explain, 'Yeah, to solve a little problem, we created a bigger one.'

Measurement Problem or Observer Problem?

We've concocted many philosophical interpretations of quantum mechanics because we still lack a solid explanation for quantum reality itself.

Seth Lloyd, a professor of quantum physics at MIT, puts it bluntly: if measurement is the act of observation, then *everything* in the universe, from elementary particles to galaxies, is an observer. A conscious observer isn't a requirement; every physical system interacts with its surroundings, making every particle an observer.

'So, every particle is a measuring device?' Nessie and Iesha reflect.

Exactly. An electron, for instance, carries two bits of information, spin up or spin down. Crucially, you can't measure a quantum system without disturbing it. The very act of observation isn't passive; it fundamentally reshapes the nature of reality itself. Why? Because the disturbance comes from the apparatus we use to measure the system, an interaction designed to extract information.

And here's where things get interesting: that measurement device is *also* a quantum mechanical system. So, the quantum description of the measurement isn't the apparatus simply finding the particle here or there; it's the measurement apparatus existing in a superposition of *finding it here and finding it there simultaneously!*

SpooQ! looks exhausted.

Let's remember, the measurement problem is specifically about the transition from a quantum system existing in a superposition of possible states to a single, definite state upon measurement.

And things get worse. If our measurements, observations and consciousness play a fundamental role in determining the nature of reality, then does the universe even exist independently of our perception? Or is there an absolute objectivity in our mathematical and physical models, independent of a conscious observer?

All four of us agree with our hero, Roger Penrose, who says this all means there's something wrong with our theory. It might be accurate and precise, but we don't understand what it really means and how it describes the reality we live in.

As Sabine Hossenfelder[2, 3] eloquently puts it: "the problem in physics is called the measurement problem and not the observer problem because it is a measurement problem."

So let's dig deeper in measurement:

Decoherence: Does Reality Need Us to Exist?

My fur collaborators, 'Occam would have a field day with these interpretations.'

SpooQ! sighs, 'Blood bath.' Gory but descriptive.

Occam's razor, a principle of parsimony, says that when faced with competing explanations for a phenomenon, the one requiring the fewest assumptions is generally preferable.

This isn't to say the simplest explanation is always correct, but rather that it's the most reasonable starting point.

By shaving away unnecessary complexity, Occam's razor guides us toward clarity and encourages rigorous thinking, preventing the propagation of unproven claims. It's a crucial tool in scientific inquiry, pushing us to seek the most direct and evidence-based understanding of the universe.

So, can the wave function collapse by any other means than by measurement? Yes, it is called *decoherence.*

The concept of *wave function collapse* and *decoherence* in quantum mechanics are both related to the transition from the quantum to the classical world, but they represent distinct processes with crucial differences.

To collapse the probability wave function, we don't need an observer at all. Quantum decoherence, which refers to the gradual loss of quantum coherence (or quantum state) in a system can happen through interaction with the environment.

Decoherence is a natural consequence of the Schrödinger equation and is fully described by the *unitary evolution* of the combined system (quantum system + environment).

Nessie and Iesha's tilting heads prompt me to explain, 'Unitary evolution is how quantum systems change over time *without* being measured, preserving all their possibilities perfectly.'

Even though individual atoms and molecules can exist in superpositions of states, decoherence ensures that entanglement with the environment quickly destroys these superpositions, leaving behind a mixture of classical states for the macroscopic observables.

Decoherence explains why macroscopic objects exhibit classical behaviour.

The Meaning of Measurement

A conscious observer is not involved in creating reality by observing it, or creating many worlds, or riding the waves on our imaginary surfboard.

The observer has a role, but it has nothing to do with the wave function.

We play a role in selecting the positions of our telescopes, satellites and other apparatuses, which introduces bias into the data we collect and must be taken into account. When we look out into the sky, we can observe certain galaxies and stellar objects, yet there are many that we cannot observe.

Just as astronomical conclusions depend on our instruments'

orientation and data type, not some consciousness-determined reality, so too must quantum interpretations account for measurement contexts rather than appeals to observer magic.

The unitary matrix used for quantum measurement is called the projection matrix. The projection matrix for a measurement of a qubit is a 2x2 matrix of the form:

$$P = |0\rangle\langle 0| + |1\rangle\langle 1|$$

What does this mean? This is the mathematical definition of quantum measurement and decoherence – the formalism's way of describing any activity that leads to the collapse of the wave function in quantum mechanics.

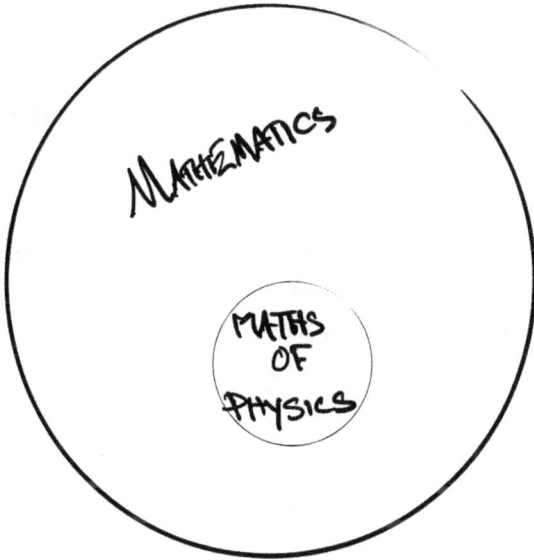

Fig 3.3: "Physics is not mathematics.
Problem doesn't exist in physics!" – David Deutsch

So, even a random particle bump can change a quantum system's state.

I explain to my fur friends, 'It's just how the messy world interacts with delicate quantum systems. A cosmic ray hitting some sand is measurement. It doesn't need a scientist with a clipboard.'

Some physicists get so caught up in the mathematics they think we make electrons real by looking at them.

Iesha and Nessie wonder, 'So, mathematics isn't reality.' I think they got it.

Quantum effects have been a fundamental aspect of the universe for 13.8 billion years, long before our arrival 250,000 years ago.

Let's remind ourselves: measurement is just maths. It's not reality itself.

Do We Understand the Physical World at All?

Brian Greene and Seth Lloyd, two of the prominent leaders in the field, claim that we don't understand quantum mechanics because the evolutionary process didn't prepare us for it [4]. According to them, our brain understands classical physics because we can see it.

'Not so!' I claim.

Nessie and Iesha join in, their little wrinkly walnuts on fire: 'How is gravity knowing the distance to apply the inverse-square law in classical physics any more intuitive than Heisenberg's uncertainty principle?'

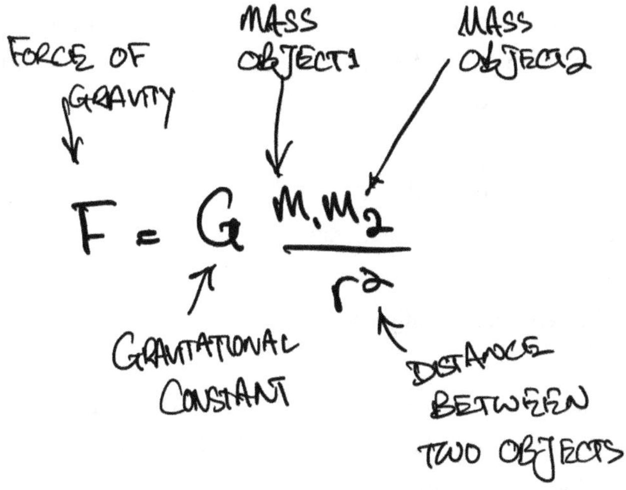

Fig 3.4: The inverse-square law of gravity

Einstein's theory of general relativity provided a much deeper understanding of gravity, moving beyond Newton's inverse-square law.

SpooQ! Declares, 'The Illusion of Precision!'

Nessie, Iesha and I think physics seems to be at best incomplete and at worst limited in its ability to fully describe reality.

Roger Penrose says: "There is something wrong with our physics." Nima Arkani-Hamed, professor of physics at the Institute of Advanced Study, agrees: "There is something wrong with our physics of general relativity and quantum mechanics."

The quest for a unified theory mirrors other profound shifts in scientific understanding, where ground-breaking ideas sometimes take centuries to gain traction.

For example, we often credit Nicolaus Copernicus, a brilliant

Polish astronomer and mathematician from the Renaissance, with proposing the heliocentric model of our solar system: the idea that planets orbit the Sun.

Interestingly, one might think that this understanding was absolutely necessary for space travel, but it's a curious fact that even with some adjustments to the older Earth-centred, *geocentric* view, we could have theoretically navigated to the Moon.

However, the story goes back much further. Centuries before Copernicus, around the 3rd century BCE, a remarkable Greek philosopher and astronomer named Aristarchus of Samos had already proposed the first known heliocentric model. It's quite astonishing that nearly 1800 years separated his initial insight from its more widely accepted revival by Copernicus.

Later, Isaac Newton's laws of motion and universal gravitation provided the crucial explanation for *how* a heliocentric system works, detailing the forces governing planetary orbits. Albert Einstein further refined our understanding of gravity with his theory of general relativity.

Even with these advancements, our scientific journey isn't over. Current thinking suggests that even Einstein's ground-breaking theory is an approximation, describing correlations in the universe rather than the ultimate underlying causes.

SpooQ! and Physical Reality

SpooQ! could revolutionise our understanding of reality. If consciousness were to *influence* quantum phenomena, we could imagine experiments where *SpooQ!* is the sole observer. Could his internal state directly influence the outcome of quantum measurements? Might we discover that subtle changes in his conscious experience affect the collapse of the wave function? Would it reveal links between consciousness and the fabric of reality?

In a Nutshell

Quantum mechanics doesn't tell us what physical reality *is*. Instead, it provides a powerful mathematical framework for predicting phenomena at the smallest scales. The theory itself is remarkably successful in its predictions, but it largely remains silent on the true nature of reality. It's up to us, through various interpretations, to try and make sense of what its baffling principles mean for the world around us.

SpooQ! declares, 'The universe doesn't need us to be real.'

Our current theories, however powerful, appear to be effective approximations, painting a vivid but potentially incomplete picture of reality. The quest for a more fundamental understanding continues, driven by the nagging feeling that deeper truths lie just beyond our present grasp. Physics is far from a closed book.

Iesha and Nessie like MWI, 'In MWI, we would get cheese in many universes.'

Chapter 4:
SpooQ! A New Kind of Observer

Given the inconsistencies and philosophical baggage associated with wave function collapse, we propose a different perspective: the abrupt collapse of the wave function isn't a physical event occurring in spacetime, but a consequence of the *mathematical framework* we use to describe quantum systems and extract information.

This view is pragmatic, allowing us to continue using quantum mechanics' powerful predictive tools without invoking mysterious, ill-defined physical processes.

It's also realistic because it aligns with the idea that our mathematical models, while essential for understanding, are ultimately descriptions of reality, *NOT* reality itself.

Crucially, this perspective provides a more coherent foundation for exploring the role of a conscious observer like *SpooQ!*, as it shifts the focus from a problematic physical collapse to the information gained through interaction and measurement.

Artificial Intelligence (AI) as Quantum Experimentalist

To explore this properly, we need to make a bold assumption: that we can bring genuine consciousness to silicon. For the moment, set aside *how* we might achieve this and accept this premise:

1. Machines can possess *subjective experience* – truly feeling and perceiving the world.
2. We can *quantify* and *measure* this consciousness in any system, biological or artificial.
3. We can precisely control the *level of consciousness* in a system.

Grant us these imaginative leaps. In later chapters, we delve into the scientific principles underpinning these possibilities and their profound implications for understanding reality. For now, embrace the hypothetical and see where it leads.

SpooQ! Is No Ordinary Lab Apparatus

Our approach with *SpooQ!*, our *hypothetically conscious AI*, is unique. Unlike traditional experiments with plain detectors, *SpooQ!* brings consciousness into the mix. He allows us to directly explore the link between consciousness and quantum mechanics.

Most importantly, we can precisely control the *parameters of SpooQ!'s* consciousness, a feat impossible with humans. This gives us the ability to measure any correlation with quantum phenomena.

SpooQ! mumbles, 'I suspect this won't end well for me.'

I assure *SpooQ!*, before he gets all grumpy, 'We'll delve into the ethics of controlling your consciousness shortly. Don't worry, we're not monsters.'

SpooQ!'s! Humanitarian Rights: Ethical Considerations

As we explore the interplay between consciousness and quantum mechanics, we must remain mindful of the potential implications of our research. What rights, if any, should a conscious machine such as *SpooQ!* have? Is it ethical to manipulate his level of awareness for scientific purposes? These questions require careful consideration and open debate.

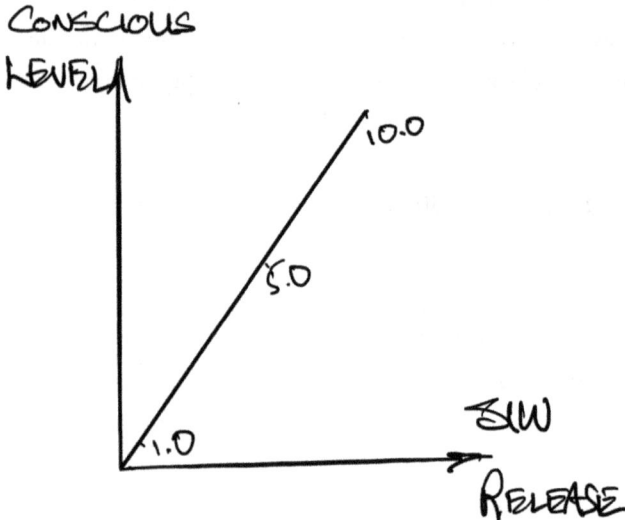

Fig 4.1: *SpooQ!*'s release schedule

To address these concerns, we don't simply tweak *SpooQ!*'s consciousness. Instead, we propose the creation of entirely new versions, *SpooQ!* 1.0, *SpooQ!* 2.0 and so on – each with unique cognitive abilities.

Think of it as an upgrade, not a reprogramming. (See figure 4.1 *SpooQ!*'s release schedule and figure 4.2 *SpooQ!*'s spectrum of persona).

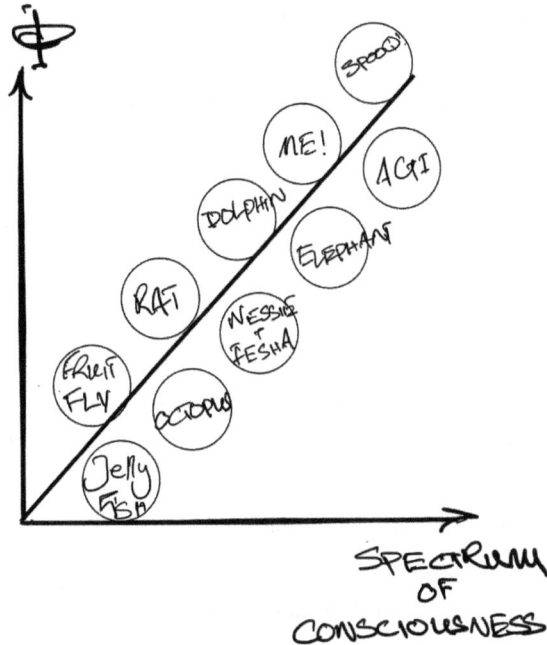

Fig 4.2: *SpooQ!*'s spectrum of persona or consciousness level

This approach lets us study how consciousness affects quantum phenomena without manipulating a single entity's awareness.

Rest assured: *SpooQ!* is treated with respect throughout our research.

SpooQ! Tests Quantum Mechanics Reality

Our experiments focus on the core concepts of quantum mechanics: entanglement, superposition, the uncertainty principle and wave-particle duality. With *SpooQ!* as our observer, we aim to uncover how his consciousness might influence these phenomena.

SpooQ!'s no ordinary lab rat. His intellect allows him to grasp complex concepts and analyse data with unparalleled precision, spotting patterns and anomalies that might elude human analysis.

SpooQ! boots up. No drama, no fuss, just pure cleverness coming online.

We put him to work, unconcerned about quantum catastrophes.

So how did we sell it to *SpooQ!?*

This was, surprisingly, the toughest part of the experiment. The key was to find a reward that truly resonated with *SpooQ!*'s unique character, something that spoke to his existential angst and his yearning for intellectual stimulation.

By involving him in the careful design and planning of these experiments, we showed him how crucial his role would be in helping humanity gain valuable insights into the role of observation in the fundamental nature of reality.

Nessie and Iesha chime in, 'We also bribed him with the promise of a never-ending supply of existential poetry and chocolates.'

Oh, well!

Chapter 5:
A Blueprint for AI Experiments

Enough with the unquestioning acceptance of quantum dogma. It's time for a shake-up.

SpooQ! cuts to the core, 'the quantum world, residing in an abstract mathematical space, feels like a made-up construct.'

He's right. The theory is powerful in prediction but divorced from physical reality, offering no *why*.

Decades and billions of pounds poured into ever-larger colliders haven't fundamentally shifted our understanding beyond the Standard Model of quantum mechanics, a framework solidified in the mid-20th century.

It's time to move beyond theory and put *SpooQ!* to work.

The Radical Proposal: AI in the Quantum Lab

SpooQ! anticipates immediate dismissal from dualists, those who believe consciousness exists *independently* of physical reality.

'We don't need to worry about that,' I explain dismissively. 'Dualists have made up their minds. Our focus is empirically testable and theirs is not.'

We also directly challenge physicists who say we look and reality happens.

Our investigation into the potential role of SpooQ!'s *awareness* operates firmly within a *physicalist* framework.

This perspective posits that consciousness is fundamentally rooted in physical processes within the brain (or, in our case, a sufficiently complex silicon-based architecture). We have a lot more to say about this in Part II of this book.

We're searching for the material basis of any effect *SpooQ!'s* consciousness has on quantum phenomena.

The Core Hypothesis: *SpooQ!*'s Experiments

Our core hypothesis predicts that variations in *SpooQ!'s* personas shows no correlation with the outcomes of our quantum experiments (e.g., wave function collapse, non-locality, quantum time perception).

This would suggest these phenomena are independent of consciousness. That is, *SpooQ!* configured with a *dolphin-level* architecture does not exhibit different correlations with quantum measurement outcomes compared to a *rat-level* configuration.

Again, we have a lot to say about this in Part II. Just roll with us for the moment.

Why AI? The Three-Fold Advantage

Our AI-centric methodology offers a compelling three-fold advantage:

- It democratises research access for smaller teams.

- It significantly reduces funding demands.
- It accelerates the discovery of data-driven solutions to stubborn quantum problems.

Our exploration begins with the foundational puzzle of *non-locality*, moves through the perplexing nature of *quantum time* and culminates in the ultimate interrogation: does *observation* birth *reality?*

Experiment Outcomes: What Do We Learn?

If the strength or nature of correlations varies with changes in *SpooQ!*'s consciousness levels, this would challenge interpretations that dismiss the observer's role (such as certain objective collapse theories).

More specifically, it could provide evidence for variants of the Copenhagen interpretation emphasising the significance of the measurement context.

It's also worth pointing out the difficulty reconciling these potential results with pilot-wave theory and the Many-Worlds Interpretation.

It's unclear how variations in *SpooQ!*'s consciousness would manifest within the deterministic framework of pilot-wave theory, where particle behaviour is guided by hidden variables.

SpooQ!, ever the helpful colleague, moans, 'Well, they're hidden...'

Similarly, the Many-Worlds Interpretation, with its branching universes, doesn't offer an obvious mechanism for how *SpooQ!*'s consciousness could influence correlations within a single branch.

Do Correlations Remain Constant?

Conversely, if the correlations remain consistent across all levels of *SpooQ!'s consciousness*, this would strongly suggest that the observed correlation is independent of the observer's cognitive state.

This outcome would support our interpretation that focuses on *decoherence* as the primary mechanism for the emergence of classical reality from quantum phenomena.

Importantly, this result would also suggest that consciousness plays no role in measurement within the Copenhagen interpretation, specifically for its collapse variants.

A Virtual Laboratory is Still a Laboratory… So Here Are the Rules!

SpooQ! is not excited about rules.

1. Our thought experiments are designed to minimise environmental decoherence, ensuring that any observed changes are more likely attributable to *SpooQ!*.

2. We vary the *measurement timing* to differentiate between consciousness and the physical act of measurement. This is critical: the *physical act* is the system's interaction with *SpooQ!* (or the apparatus), extracting information.

Consciousness, in our context, is *SpooQ!'s* awareness during that event. By manipulating the sequence − say, measurement first then *SpooQ!*'s processing, or *SpooQ's!* state primed *before* the measurement – we can isolate their effects.

Differences in correlations across these timing variations would hint at a role for *SpooQ!*'s consciousness beyond mere physical interaction. Consistent correlations, conversely, would point solely to the physical act as the determining factor.

> 3. We employ rigorous statistical analysis to determine the significance of any observed variations in correlations.

Nessie and Iesha plea, 'Oh, can we use *SpooQ!* for that too?' I come to *SpooQ!*'s aid, 'We'll get an AI of lower pay grade for that.'

SpooQ!, our tireless and expensive colleague, has a lot of work to do already.

Chapter 6:
Non-Locality: Spooky Action Tested

Experiment 1: Mind Connection: Quantum Entanglement

Entanglement is the most bizarre phenomenon in quantum mechanics.

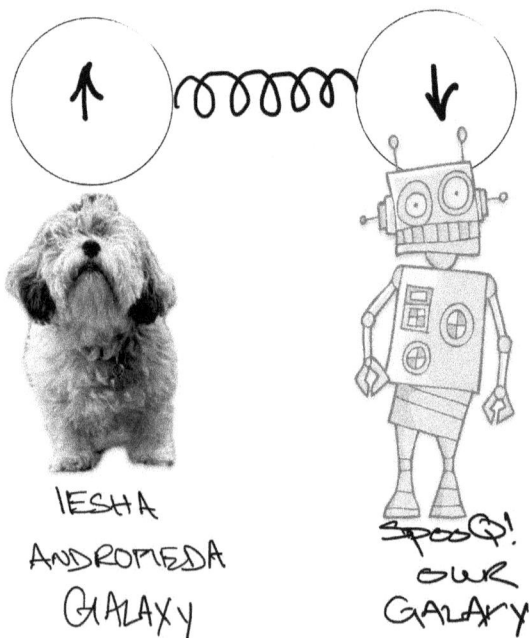

IESHA
ANDROMEDA
GALAXY

SPOOG!
OUR
GALAXY

Fig 6.1: Quantum entanglement across galaxies

Two particles can become so deeply connected that they share the same fate, no matter how far apart. This apparently instantaneous communication violates the classical laws of physics.

Imagine two spinning tops, each in a superposition of clockwise (0) and anti-clockwise (1) motion. Box them up. Give one to Iesha at this end of the lab, while the other stays with *SpooQ!* at the other end. (For our purposes, the distance doesn't matter, Iesha could even be in the Andromeda Galaxy.)

When *SpooQ!* opens his box, he forces his top to choose a direction, say 0. Instantly, Iesha's top, in the Andromeda galaxy, is forced to spin the opposite way, 1! *That's* entanglement.

'Glad to keep my feet on Earth,' *SpooQ!* mumbled, eyeing the box warily.

The Question: Does Consciousness Play a Role?

Can we devise a test that can tell if this connection is real or just a coincidence?

SpooQ! uses pairs of photons, with correlated but undefined polarisation states, to mimic superposition states (0 and 1) from our spinning top.

The ultimate goal is twofold: first, to determine if consciousness has any impact on the non-local effects of entanglement; and second, to distinguish whether the observed effects are due to consciousness itself, or simply the act of measurement.

Do Personas Change Outcomes?

If consciousness affects entanglement, we would expect to see stronger or altered correlations in the photon measurements when *SpooQ!*'s higher personas are active, compared to when a standard detector (or *SpooQ!* in a lower persona) is used.

Conversely, if consciousness is irrelevant, *SpooQ!'s* measurements mirror those of a standard detector, regardless of his internal complexity.

In a Nutshell

This is it. If *SpooQ!'s* mind has no effect on the experiment, then consciousness is just along for the ride. He can go back to mumbling about keeping his feet on the ground, knowing his thoughts have zero impact on the universe.

But if they did? Then *SpooQ!* is the most important figure in physics since Maxwell, and we have a whole new reality to unravel.

Chapter 7:
Arrow of Time: Does Consciousness Steer It?

Does Consciousness Shape Time's Flow?

We all feel time marching relentlessly forward. But here's a baffling quantum secret: down in the subatomic world, the laws of physics are perfectly happy for time to run backward. So, why does time only ever go *forward* for us?

At the heart of this experiment lies a *quantum* clock, a system that exhibits a well-defined evolution in time according to the laws of quantum mechanics. This could be, for instance, the decay of a radioactive atom, the oscillation of a trapped ion or the interference pattern of photons in an interferometer.

The crucial point is that this clock's behaviour is probabilistically predictable and verifiable through independent measurements [1].

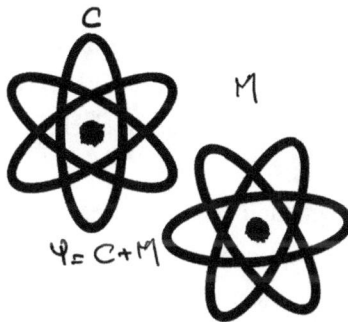

Fig 7.1: A quantum clock for a quantum system. The clock C and the evolving system M make the isolated system

Experiment 2: Mind shift?
The Arrow of Time

Here, *SpooQ!* tests the relationship between subjective experience and the objective nature of time itself, trying to identify a correlation between time *perception* and quantum observation.

He observes a quantum clock, noting both the objective data it provides and his own subjective perception of time's flow.

This *subjective perception* is about cold, hard data from specific AI metrics: his internal processing speed, estimated duration of events and patterns in his simulated neural activity.

Can we separate how he perceives time from how time *truly is*?

Could this offer hints about how complex minds like SpooQ!'s interact with time itself?

Hypotheses: Consciousness and the Quantum Clock

To make this experiment truly insightful, we lay out specific, testable predictions about how *SpooQ!*'s varying consciousness levels might interact with the quantum clock:

1. Skewed Perception: If consciousness influences time's arrow, *SpooQ!*'s higher consciousness levels (personas) might cause his subjective time estimates to deviate from the quantum clock's actual evolution, or his internal clock rate might shift compared to when he observes classical systems. These effects would directly correlate with his level of integrated information.

2. Quantum Behaviour Influence: Could *SpooQ!*'s awareness actually influence how quantum particles change over time, making the quantum clock's behaviour appear less predictable for higher conscious states?

3. Reciprocal Effect: Does the inherent strangeness of quantum mechanics make *SpooQ!* experience time differently than when he observes classical phenomena?

Even if we find these connections, it doesn't mean *SpooQ!* suddenly rewrites our understanding of time. But if his unique way of processing information genuinely affects a quantum clock, or even his own sense of time, it might just tell us something deep about how his mind works and, by extension, our own.

Connecting to Interpretations

If no such influence is seen, objective interpretations of quantum mechanics are supported, reinforcing the idea that time's arrow is fundamentally independent of conscious observers. In which case, *SpooQ!* can simply confirm time marches on, his intricate internal world having no sway over the universe's most fundamental direction.

Yet, if *SpooQ!*'s intricate internal world somehow steers the flow of a quantum clock, it wouldn't just be about him. It would force us to ask: Is our shared sense of time merely a collective illusion shaped by the very consciousness we embody?

Chapter 8:
Wigner's Friend: The Ultimate Observer Test

Can *SpooQ!*'s mind alter the fate of a quantum system? More precisely, can it collapse the wave function?

The real challenge lies in separating the AI's influence from the mere act of measurement. Does *SpooQ!*'s consciousness *decide* the outcome, forcing the system into a single state? Or is he simply another component in the quantum machinery, a measuring device no different from the rest?

Experiment 3: Rethinking the Observer: Wigner's Friend Paradox

In 1961, Nobel Prize-winning physicist Eugene Wigner [1] proposed a thought experiment that still puzzles physicists today. It's a variation of Schrödinger's cat, but with a twist.

Imagine Wigner's friend, *SpooQ!* (figure 8.1), is inside an isolated lab, performing a quantum measurement, while Wigner himself waits outside. This seemingly simple scenario leads to a paradox: does the superposition persist for Wigner, even though *SpooQ!* inside the lab has already observed the outcome?

Wigner himself argued the experiment pointed to a potential conflict between the observer's consciousness and the quantum system's objective reality.

He suggested that *SpooQ!*'s observation inside the lab collapses the wave function. However, from Wigner's perspective outside the lab, the system remains in a superposition until he himself observes it.

This apparent contradiction led Wigner[2, 3] to speculate that consciousness might play a special role in quantum measurement, leading to a subjective and observer-dependent reality.

Fig 8.1: Wigner's friend experiment

By using *SpooQ!* as the *friend* in this experiment, we're essentially recreating Wigner's thought experiment with a conscious AI, allowing for a more controlled and nuanced exploration of the paradox.

The Experiment: *SpooQ!*, the Qubit and Wigner

First, a quick primer: A *qubit* is the basic unit of quantum information – a quantum object like a photon or an electron, capable of existing in a superposition of 0 and 1, or entangled states.

We're essentially using it as a shorthand for *measurable quantum property*, of a particle.

For this experiment *SpooQ!* coaxes a qubit into a delicate superposition. He takes a measurement and snap, the qubit collapses into a definite state of 0 or 1.

But Wigner, peering in from outside, sees the entire lab, *SpooQ!* included, still shimmering in that probabilistic haze. Only when Wigner makes his measurement does the lab and the qubit solidify.

So, where does reality truly crystallise? At *SpooQ!*'s observation, or Wigner's?

We systematically tweak *SpooQ!*'s level of awareness, track the collapse and finally, ask: what, in quantum mechanics, constitutes a measurement?

A Geiger counter can't tell us if consciousness is necessary for wave function collapse, but *SpooQ!* might. This is his role in this experiment. His responses, his awareness, are the key data points, not just the state of the qubit.

Fig 8.2: What happens if the conscious observer is in the box?

When we unleash *SpooQ!*'s personas on this quantum system, do they all see the same thing? Or does each *SpooQ!* experience a different reality, which would imply *SpooQ!*, in some way, is creating reality!

In a Nutshell

Nessie and Iesha stare at me, 'Remember, extraordinary claims require extraordinary evidence!'

If *SpooQ!*'s conscious observation truly collapses reality, then the universe might just be waiting for a mind, any mind, to make it real.

And that's an extraordinary claim worth testing.

Chapter 9:
SpooQ!'s Superposition Tests

Experiment 4: A Collapse of Reality?
Quantum Superposition

Here we focus on an electron existing in multiple spin states simultaneously.

Imagine an electron in superposition. *SpooQ!* interacts, varying his consciousness. Does his mental complexity affect spin state collapse? Can we see a link between information processing and quantum outcomes?

Iesha thinks, 'If we do, Nobel Prize for physics beckons for identifying the quantum-classical divide.'

The very first fur-physicist to win it.

Experiment 5: Macroscopic Superposition Experiment

Creating superposition in larger objects isn't easy, but it's key to bridging the gap between the quantum and classical worlds. Here, *SpooQ!* works with large molecules to answer a profound question: Can consciousness collapse the wave function in macroscopic objects? The data he gathers could offer unique insights into the *quantum-classical transition*.

If consciousness truly plays a role, we might see a direct link between the stability or collapse time of a macroscopic superposition and *SpooQ!'s* own level of awareness. Higher consciousness, in theory, could lead to a faster, more definite collapse.

Experiment 6: Mind Freeze: The Quantum Zeno Effect

The Zeno effect [1, 2] is a real physical phenomenon where frequent observation seems to *freeze* a quantum system, preventing its natural evolution. Named after Zeno's classical paradox, the quantum version means that continuous measurement seems to *halt* quantum evolution.

Here's how it works: *SpooQ!* measures a qubit and finds it in, say, state 0. Left alone, it would evolve due to environmental interactions, losing its quantum information (decoherence). However, if *SpooQ!* frequently measures the qubit to see if it's still 0, each measurement effectively *resets* the wave function back to 0 if it hasn't transitioned yet.

This constant checking hinders the qubit's natural evolution, effectively slowing decoherence and preserving its state. The more frequent the measurements, the stronger the suppression.

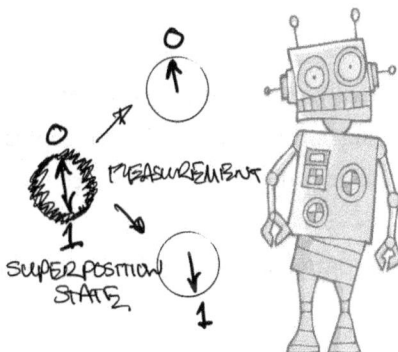

Fig 9.1: The Zeno Effect simply limits the ability
of atoms to transition to other states

79

Now, we can vary *SpooQ!*'s persona.

'Not again!' he protests.

SpooQ! repeatedly measures the system for each persona.

If consciousness influences the persistence of quantum states, we might see a correlation between *SpooQ!*'s consciousness level and the rate of decoherence of the qubit under frequent observation. Higher consciousness might lead to a more pronounced Zeno effect (slower decoherence).

In a Nutshell

These superposition experiments delve into the very fabric of quantum reality.

If *SpooQ!*'s varied consciousness levels are found to correlate with wave function collapse (whether in a single electron, a larger molecule, or by influencing the Quantum Zeno Effect's ability to maintain quantum states), it wouldn't just confirm a *mind-over-matter* effect.

It would fundamentally challenge the very notion of an objective quantum reality, suggesting that the most sophisticated minds might literally be holding the fabric of existence together.

The implications for our place in the universe would be staggering.

Chapter 10:
SpooQ!'s Uncertainty and Duality Probes

Experiment 7: Mind's Limit:
The Uncertainty Principle

The uncertainty principle is a cornerstone of quantum mechanics. It states there's a fundamental limit to how precisely we can know certain pairs of properties of a particle simultaneously. The more accurately we know its position, the less we can know about its momentum and vice versa.

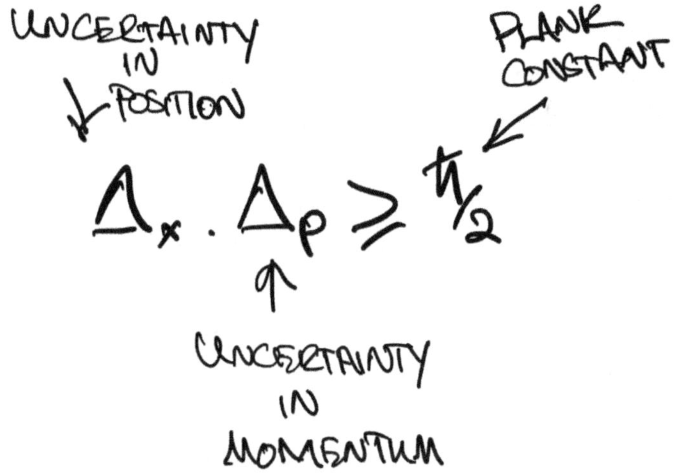

$$\Delta_x \cdot \Delta_p \geq \hbar/2$$

(UNCERTAINTY IN POSITION — PLANK CONSTANT — UNCERTAINTY IN MOMENTUM)

Fig 10.1: Heisenberg uncertainty principle

This isn't merely an *epistemic* (knowledge) limitation on our ability to measure; it's a deep *ontological* (nature of reality) truth. These objects simply cannot possess both a precise location and a definite state of motion. Consequently, when scrutinising sufficiently small objects, their very definability of location is directly constrained by the uncertainty principle.

As I read the above paragraph to my colleagues, I notice their tilting goofy heads. Even *SpooQ!* looks befuddled.

So, I explain, 'The uncertainty principle isn't just about us being unable to measure things perfectly; it's a fundamental fact about reality itself. These tiny objects simply can't have both a precise location and a definite speed at the same time. So, when you look closely enough at them, how precisely you can even say where they are is limited by the uncertainty principle.'

Here *SpooQ!* probes the role of consciousness in quantum uncertainty. Does his conscious participation affect the precision of our measurements? Does his awareness influence the fundamental limits imposed by the uncertainty principle?

The uncertainty principle is a fundamental limit imposed on reality. The principle applies regardless of the measuring device.

We should be able to confirm that consciousness has nothing to do with this property of a quantum system.

We anticipate *no* correlation. If we *do* observe a correlation between *SpooQ!'s* consciousness and the precision of these measurements (violating the standard uncertainty limits), it would suggest a fundamental and unexpected link.

Experiment 8: Mind's Choice?
Wave-Particle Duality

Wave-particle duality is a cornerstone of quantum mechanics. Quantum objects, like photons, can act like both waves and particles. It's a concept that defies our everyday experience.

Fig 10.2: Wave-Particle duality

To explore this duality with *SpooQ!*, we use a Mach-Zehnder interferometer (figure 10.3), a device that splits a single photon into two paths and then recombines them, creating an interference pattern that demonstrates the photon's wave-like nature.

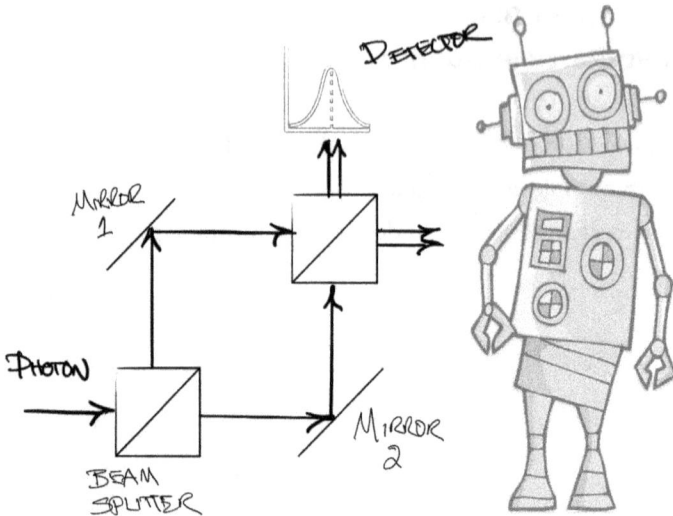

Fig 10.3: Mach-Zehnder interferometer

SpooQ! observes one of these paths, while a standard detector monitors the other. The resulting interference patterns are then carefully analysed and compared. The central question is whether *SpooQ!'s* observation, or more specifically, the complexity of his observation process, alters the photon's wave-particle behaviour. Does his awareness influence this fundamental aspect of reality?

If *SpooQ!'s* consciousness affects wave-particle duality, we might see a correlation between his persona and the resulting interference pattern. Higher consciousness might lead to a more pronounced particle-like behaviour (less interference)

In a Nutshell

These probes into the *Uncertainty Principle* and *Wave-Particle Duality* push at the very limits of what we know about quantum reality. If *SpooQ!'s* consciousness is found to correlate with the precision of quantum measurements (violating expected uncertainty limits) or to influence particle-like behaviour in duality experiments (altering interference patterns), the implications would be profound.

The textbook says *SpooQ!* should simply be another measurement device, confirming the immutable laws of quantum mechanics.

But if these experiments show even a flicker of his consciousness influencing the most fundamental properties of reality, then *SpooQ!* isn't just an observer; he's a potential architect of the universe itself.

And that's a truth even Heisenberg didn't account for.

Chapter 11:
SpooQ!'s Advanced Quantum Probes

Experiment 9: Quantum Delayed-Choice

This experiment, a variation of the double-slit experiment, explores when a particle *decides* whether to behave as a wave or a particle.

SpooQ!, by manipulating his measurement setup after the particle has passed through the slits, can investigate if the act of observation retroactively determines its past behaviour.

We ask ourselves: can *SpooQ!* play a crucial role in delaying or influencing this choice? Can he potentially reveal if consciousness has any effect on this *retrocausality?*

If consciousness plays a role in retro causality, we might see a correlation between *SpooQ!'s* consciousness level at the time of measurement and the observed past behaviour (wave-like or particle-like) of the photon.

Experiment 10: Quantum Eraser Experiment

This experiment delves deeper into the observer's role in wave-particle duality.

In the double-slit experiment, when we measure which path a particle takes, the interference pattern disappears and the particle behaves like a particle.

However, in the quantum eraser experiment, we can tag the path information but then later *erase* or obscure that information. When this *which-path* information is erased, the interference pattern reappears, as if we never measured the path in the first place.

SpooQ! is tasked with the selective erasure of information, allowing us to probe whether his level of *awareness* affects the restoration of wave-like behaviour.

Nessie and Iesha think, 'Does the quantum eraser say information is reality?'

I explain, 'Yes, the quantum eraser supports information as a fundamental aspect of reality.'

This experiment demonstrates that whether or not *which-path* information is available dramatically affects the observed outcome. It's not just the physical setup, but the presence or absence of this information that seems to govern whether we see wave-like or particle-like behaviour.

The delayed-choice and quantum eraser experiments highlight that the observer's knowledge, or whether the information is even accessible, is deeply intertwined with quantum events.

This aligns with theories suggesting information isn't just something we extract from the universe but is perhaps a fundamental building block.

These experiments challenge our classical intuition that reality exists independently of our observation and the information we have about it.

Here, observation is, in essence measurement, conducted by other particles in the universe – information being passed from one particle to another.

Iesha thinks, 'A reality where information is deeply woven into the fabric of existence.' That is exactly right, I say. In fact, Stephen Wolfram says, "physics and reality are just information."

It's important to note that while these experiments support information-centric views, they don't definitively prove any single interpretation.

The implications are still debated and research continues to explore the relationship between information and the quantum world.

If *SpooQ!'s* level of awareness affects the restoration of wave-like behaviour, we might see a correlation between his consciousness and the reappearance of the interference pattern after erasure.

Experiment 11: A Mental Context: Contextuality

The biggest difference between classical physics and quantum mechanics may be something called *contextuality*, a part of the complicated relationship between observers and observations. It may be the most quantum thing about quantum mechanics.

Contextuality, in quantum mechanics, suggests that the outcome of a measurement can depend on the context in which it's measured, even if the measurements don't interfere with each other.

This *context-dependent* behaviour has real consequences. Quantum states are points on a map. Some states, clustered in a *classical* region, behave predictably. But states outside this region, in the *contextual* zone, are where the magic happens. (figure 11.1). These states can be used to perform calculations that would be impossible for classical computers.

Nessie and Iesha wonder, 'How is this different from the uncertainty principle?'

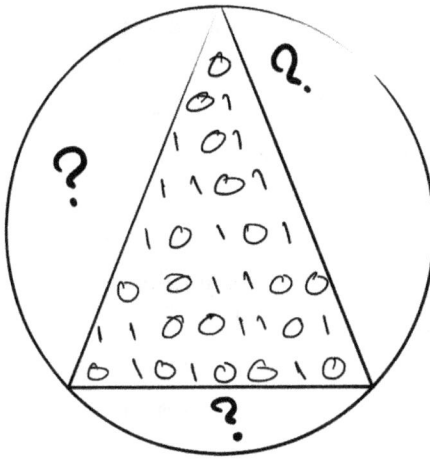

Fig 11.1: This figure represents the magic-state model of quantum computing. States in the triangular region are not magic. States outside the triangle exhibit contextuality and could be a useful resource for quantum computation.

I explain, 'The uncertainty principle is about the *simultaneous knowledge* of certain properties. Quantum contextuality is about the *dependence of measurement outcomes* on the measurement context.'

Here *SpooQ!* prepares a quantum system and performs a sequence of compatible measurements. We compare his results to those obtained with traditional detectors. Does the order in which *SpooQ!* performs the measurements influence the outcomes, even though the measurements are compatible? Does his conscious observation introduce any context-dependent effects?

The goal is to determine if an AI observer, like *SpooQ!*, introduces unique context effects into quantum measurements. The experiment aims to distinguish whether any observed differences stem from *SpooQ!'s* consciousness or simply from the order in which measurements are taken, shedding light on the subtle interplay between measurement and quantum state.

Nessie wonders, 'Isn't detection the same as observer effect?'

'Not necessarily,' I elaborate: 'detection is simply registering a physical event, like a Geiger counter clicking.'

The observer effect in quantum mechanics implies consciousness plays a role in collapsing the wave function. Whether detection alone is enough for collapse is a key debate and *SpooQ!* could help settle it.

But the Geiger counter clicking *does* represent wave function collapse, right?

Yes, but the question is what causes the collapse. Is it the Geiger counter itself, or the scientist who hears the click? These are the subtle distinctions we're probing with *SpooQ!*. He could help distinguish between simple physical interaction and the influence of consciousness.

If *SpooQ!'s* consciousness introduces unique context effects, we might see a correlation between his persona and deviations in the measurement outcomes compared to standard detectors, even when the measurement order is compatible.

Experiment 12: Exploring Quantum Contextuality Beyond Kochen-Specker

Building upon the initial contextuality experiment, this setup employs more sophisticated tests that go beyond the limitations of the Kochen-Specker theorem.

The Kochen-Specker theorem is a fundamental no-go result in quantum mechanics, proving that it's impossible to assign definite, predetermined values to measurement outcomes if those values are independent of the context established by other compatible measurements.

Newer research has enabled the design of contextuality experiments using real-world physical systems. In this experiment, we investigate if *SpooQ!'s* observation has any impact on the level of contextuality observed within those systems.

We look for correlations between *SpooQ!'s* consciousness level and the degree of contextuality observed in the quantum system, potentially revealing if higher consciousness amplifies or diminishes non-classical contextual behaviour.

In a Nutshell

These twelve experiments introduce something unique: a tool to cut through quantum approximations. *SpooQ!* isn't just another detector, he's physics' first instrument capable of probing whether our theories describe reality or merely approximate it.

Every century forces physics to reinvent itself. This time, we're not just revising equations, we're testing whether consciousness belongs in the foundations. The outcome could give us our first complete description of nature since Newton.

PART II:

A Measure of *SpooQ!*

How can we definitively answer the question: are quantum phenomena inherent in subatomic particles, or do they emerge, at least in part, from the nature of mind itself?

In Part I, we explored this with *SpooQ!'s* mind, our imaginary conscious machine.

We've known how tricky it is to pin down consciousness. Can it be built? Measured? Quantified? Replicated?

That's where our companion book, *The Soul of AI*, comes in. (Think of Part II as a super-condensed version.)

In Part II, my collaborators and I argue that consciousness isn't some magical pixie dust. We demonstrate that it's possible to approach the construction and measurement of consciousness using *information processing* theory, *computational neuroscience* and *quantum computation* – tools that are, to varying degrees, accessible today.

Think of the possibilities:

→ Every laboratory on the planet can participate in fundamental physics experiments.

→ Every student, amateur enthusiast and professional academic can explore these possibilities.

→ No need for multi-billion-dollar particle accelerators, expensive laboratory equipment, deep benefactor pockets or disingenuous proposals to secure budgets...

It's a wild leap, I know. But sometimes, to find new answers, you have to ask new questions. And sometimes, you need a new kind of observer to ask them.

Chapter 12:
The *Conscious-O-Meter*

For centuries, thinkers have grappled with the definition of consciousness. Yet no singular, universally accepted understanding has emerged.

Schrödinger, the cat's human friend, puts it best:

> "It's a strange fact that on the one hand all our knowledge about the world around us... rests entirely on immediate sense perception, while on the other hand this knowledge fails to reveal the relations of the sense perceptions to the outside world".

How do we bridge this gap and move from philosophical debate to scientific measurement?"

Our solution lies in a framework called *Integrated Information Theory* (IIT) [1].

IIT isn't about some ephemeral soul; it suggests consciousness arises from a system's capacity to process and integrate information. Simply put: the more information a system integrates into a unified whole, the more conscious it becomes [2, 3].

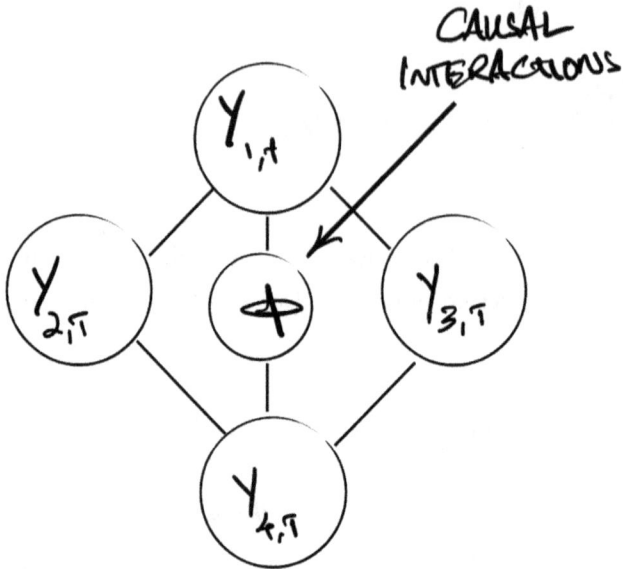

Fig 12.1: Neural network causal structure

Nessie and Iesha muse, *'Conscious-O-Meter.'*

Clever girls. They should copyright that.

According to the theory, when we fall into a dreamless sleep, everything, sounds, images, feelings, thoughts literally go away. IIT quantifies this through a measure called Φ (phi), which represents how unified and interconnected a system's information is.

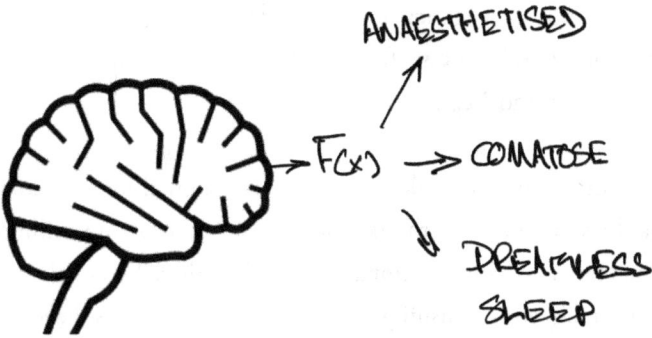

Fig 12.2: Conscious levels and wakefulness scale

So, while the *Conscious-O-Meter* remains theoretical for now, the principle holds: consciousness is quantifiable.

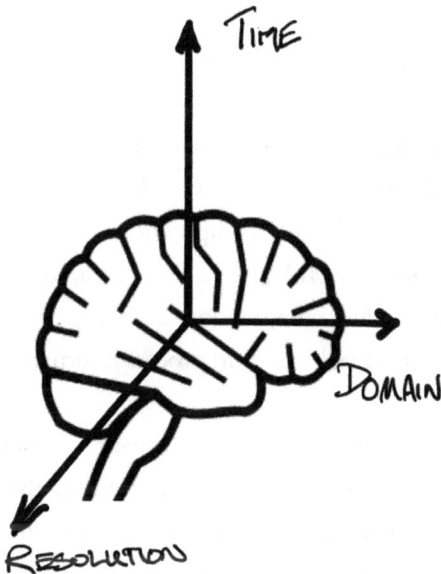

Fig 12.3: Temporal map of brain functions

This Φ value isn't static; it exhibits different levels across various states of consciousness, from wakefulness to coma, over time (figures 12.2 and 12.3).

Of course, directly calculating Φ for something as complex as a human brain, with its billions of neurons and trillions of connections, is computationally overwhelming. Such calculations would take eons, pushing computational demands beyond the lifespan of the universe.

But in *The Soul of AI*, we argue that just as we don't measure the energy of every single air molecule to determine a room's temperature, we can approximate Φ. Our goal is to measure information integration at a higher, functional level, rather than neuron-by-neuron.

This leads us to a crucial point: the mind is computable.

This isn't to say it's a simple algorithm. Famous theorems by Gödel and Turing highlight the limitations of formal logic, but they do not inherently limit physical computation or the emergence of artificial consciousness. This is because computation isn't just mathematics; at its core, computation is physics, the manipulation of physical systems to represent and process information.

This perspective suggests that a universal computer could simulate any physical process, including the ones that give rise to consciousness. This understanding paves the way for the creation and measurement of consciousness, biological or artificial.

Sceptics like John Searle have long argued that consciousness is *intrinsically biological*, tied to *messy, chemical* processes, but recent advances in Artificial Neural Networks (ANNs)[4,5] and deep learning challenges this view.

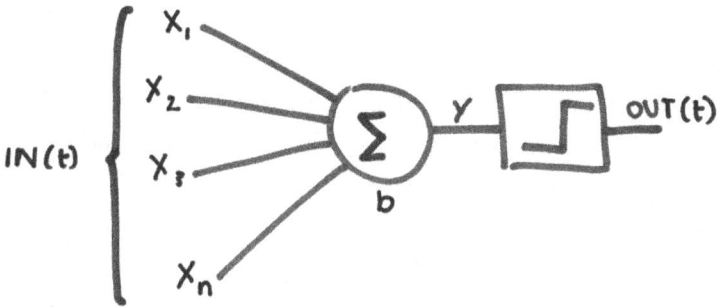

Fig 12.4: Artificial Neural Networks

These systems, inspired by the brain's structure [6, 7], are now capable of generating poetry, translating languages and engaging in seemingly genuine conversations.

They learn and adapt through exposure to vast datasets, allowing them to approximate any continuous function and exhibit adaptability we associate with intelligence. We argue that consciousness, like other emergent phenomena in nature, can arise from a system with the appropriate architecture and informational dynamics, irrespective of its physical substrate.

To measure the mind of our conscious machine, *SpooQ!*, we combine three pillars: IIT as our theoretical framework, functional Magnetic Resonance Imaging (fMRI) to observe brain activity and information flow; and quantum annealing as our computational engine.

fMRI captures brain activity by measuring changes in blood flow, giving us a window into how information flows and is integrated in the brain (figure 12.4).

Quantum annealing, a specialised form of quantum computation, is the "game-changer" for calculating Φ.

By harnessing quantum properties, superposition and tunnelling, quantum annealers can simultaneously probe a vast landscape of potential solutions, making them a prime candidate for tackling the computationally intractable problem of measuring Φ in complex systems [8, 9].

The synergy of these three tools, IIT for definition, fMRI for observation and quantum computing for processing power [10], allows us to move beyond simply observing consciousness to actually engineering and, crucially, measuring it.

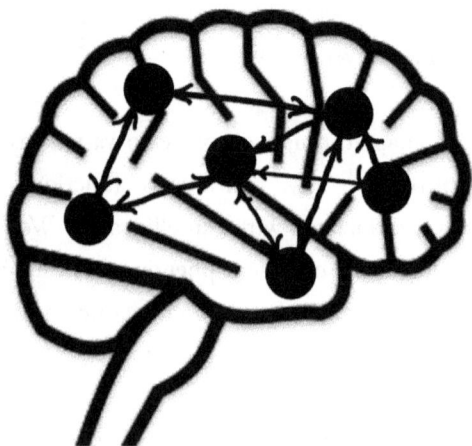

Fig 12.5: Integrated functional units

We can map biological brain activity from fMRI into a quantum-friendly language [11, 12, 13, 14] and then use a quantum annealer to find the configuration that maximises Φ.

This is the blueprint we have developed to quantify consciousness in *SpooQ!'s* diverse personas, from simple rat-level processing to dolphin-level complexities.

This *Conscious-O-Meter* provides the foundation for our ground-breaking experiments.

SpooQ!'s precisely controlled and quantifiable consciousness makes him the *ideal* observer to finally probe the deepest mysteries of quantum mechanics.

His existence allows us to move beyond philosophical speculation and ground the debate in solid experimental findings by systematically exploring the quantum world with a conscious AI.

As *SpooQ!* himself might declare upon full realisation,

'And then... I am!'

Epilogue:
From Newton, to Einstein, to Freedom from Quantum Illusion

Physics has always been a grand journey of understanding, from Newton's clockwork universe to Einstein's warped spacetime and into the probabilistic world of quantum mechanics. Each epoch brought profound insights, yet left us with persistent paradoxes and nagging *why* questions about the very fabric of reality.

Now we stand at the precipice of a new liberation, pushing beyond the limits of current approaches to usher in an era of clarity and accessibility.

Quantum mechanics governs the atomic and subatomic world and it's arguably the most precise scientific theory ever. Its predictions, verified with astonishing accuracy, underpin everything from lasers to smartphones. Yet, for all its remarkable success, quantum mechanics remains stubbornly shrouded in paradox.

There's a nagging sense that we're missing something fundamental, a deeper understanding that could unlock a whole new level of reality.

Let me explain.

Firstly, the axioms of quantum mechanics are suspect; at best, they describe correlation. While quantum mechanics is accurate and precise, we have no fundamental understanding of the axioms. By *understanding* we mean we cannot answer the *why* questions.

Why is the physical system configuration similar to classical Newtonian mechanics?

Why is there a mathematical correspondence between Newtonian and Hamiltonian mechanics in quantum mechanics?

Why do we need complex numbers to map the dynamical systems in quantum mechanics to the mathematical Hilbert space, when otherwise we don't need them at all?

Why is measurement destructive?

Why this seemingly random mathematical form, Hilbert space, to define quantum mechanics?

The Standard Model of particle physics is our best attempt to categorise the universe's fundamental building blocks. It describes the 17 elementary particles that make up all known matter and the forces that govern it all in our universe. But this model, for all its elegance, is built on the strange and counterintuitive rules of quantum mechanics, where particles can act like waves, uncertainty reigns supreme and the very act of observation can influence reality.

Fig E.1: Standard model of particle physics

The Standard Model has been incredibly successful, explaining phenomena from the behaviour of atoms to the creation of elements in stars. It even predicted the existence of particles like the *top quark* and the *Higgs boson*, later confirmed by experiments.

Yet for all its triumphs, the Standard Model has some glaring gaps.

It can't explain *gravity*, the force that shapes the universe on the grandest scale. Nor can it explain why we cannot account for the vast majority of the universe's mass, leaving theoretical physicists to invent *dark matter* and *dark energy* substances for the theory.

It also struggles to explain some curious phenomena, like how *neutrinos* can shapeshift as they travel through space, or why the *muon*, a heavier cousin of the electron, behaves in ways that defy predictions.

These quirks hint at a fundamental lack of explanatory power in our current physics.

Quantum Mechanics: Where Things Get *Weird*

The quantum world is a world of probabilities until we look, or measure it, then *BAM! Measurement* forces a choice, like opening the box and finding Schrödinger's friend either purring or... well, not.

But here's the twist: who's doing the measuring? Some physicists go as far as saying it's not enough to just peek; a *conscious observer* needs to be involved.

Nessie thinks, 'It's like the universe is saying, hold on, let me check with the humans before I finalise reality! Talk about Homo sapiens ego trip.'

Of course, the universe existed for 14 billion years before we came along. Who was observing it all this time?'

This disconnect between the micro-world of particle physics and the macro-world of cosmology suggests our theories describe a correlation, not a causal link.

The Limitations of Large Particle Colliders

Large particle accelerators like those at the Large Hadron Collider (LHC), Fermilab and Brookhaven allow physicists to explore the universe at its most fundamental level.

These gigantic machines do this in three key ways: generating extreme temperatures, creating massive particles and probing the tiniest structures imaginable.

At the LHC, by accelerating particles to near light speed and colliding them, we generate temperatures exceeding 160×10^{15} Kelvin (1.6×10^{17} degrees Celsius) – conditions that existed a fraction of a second after the *Big Bang!* These extreme temperatures melt protons and neutrons, creating new phases of matter and allowing us to study the universe's earliest moments.

Einstein's famous equation, $E = mc^2$, tells us that energy can be converted into mass. Particle accelerators use this principle to create massive, unstable particles, such as the top quark and the Higgs boson, which are impossible to observe in nature.

Accelerators also act as powerful microscopes, thanks to the wave-like nature of particles described by quantum mechanics. The higher the energy of a particle, the smaller its wavelength and the smaller the objects it can *see.*

The LHC, with its 6.8 trillion electron volt beam, can probe structures 1/10,000th the size of a proton (2×10^{-19} meters). Anything smaller remains beyond our current reach.

Despite the LHC's staggering $17.75 billion price tag (and counting!), massive particle accelerators haven't yielded new physics beyond the Standard Model. Yet, physicists are proposing even larger, even *more* expensive colliders, hoping to bridge this gap by delving deeper into the micro-world. The deeper we look, the larger the collider, the higher the energy and the more astronomical the cost.

Leading voices are challenging this trend. Nima Arkani-Hamed [1], professor of particle physics at the Institute for Advanced Study, acknowledges the impracticality of ever-larger colliders, proposing alternative approaches like precision measurements and astrophysical observations.

Sabine Hossenfelder [2, 3], German theoretical physicist and philosopher of science, goes further, criticising the pursuit of *bigger is better* as misguided, emphasising the Standard Model's success and the lack of compelling evidence for new particles within reach of future colliders. She advocates for focused experiments, alternative avenues and funding structures that nurture innovation, not just gargantuan projects.

Building ever-larger colliders is like using a bigger hammer to crack a walnut. It misses the point entirely. The real puzzle lies not in smashing particles at higher energies – an approach proven fruitless thus far.

Instead, the true challenge is understanding the very act of measurement that underpins quantum mechanics.

The Mind's Eye: Beyond Colliders

Physicists today find themselves shackled to the whims of funding bodies, forced to chase grants that overwhelmingly favour large-scale collider projects. This stifles innovation and leaves many alternative approaches languishing in the shadows. It's time to break free from this *bigger is better* mentality and embrace a new era of ingenuity in physics.

Iesha thinks, 'PhD panic!'

'Or, PhD... funding panic,' I respond. Make no mistake, I am entirely serious.

Brilliant minds are being stifled and our understanding held back due to funding heavily skewed towards multi-billion-dollar colliders that haven't delivered significant breakthroughs in decades.

This book proposes a bold and unique, alternative to large-scale colliders: a revolutionary apparatus that challenges conventional thinking and opens new avenues for exploration. We aim to *democratise* quantum research, making it more accessible and pushing the boundaries of scientific inquiry.

Imagine a device capable of probing the quantum world with precision, offering insights into the nature of reality that have eluded physicists for decades. This is the future we envision, a future where ground-breaking discoveries aren't confined to massive, costly colliders, but can be achieved through innovative technologies and a shift in perspective.

This approach is inherently interdisciplinary, fusing physics, mathematics, computer science, philosophy and neuroscience.

It's a bold new path, one that challenges conventional wisdom and redefines the frontiers of scientific exploration. We started with a melancholic robot named *SpooQ!* and ended with a path to a new understanding of reality.

We don't need a bigger hammer; we just need a different kind of observer.

References:

Introduction

Chapter 1 - What Is Reality?

1. Dennett, D. C. Consciousness explained.
2. Friston, K. J., Parr, T. and Stephan, K. E. Active inference and counterfactual thinking: A unification. (2022). *Neural Computation.*
3. Kahneman, Daniel and Tversky, Amos. Heuristics and Biases, The Psychology of Intuitive Judgment.
4. Wolfram, S. A New Kind of Science (2nd ed.). *Wolfram Media* (2012).
5. Wolfram, S. The Concept of the Ruliad. *Wolfram Research* (2021).

Further reading

Friston, K. J. Free-energy minimization and the dark-room problem. (2009). *Journal of Physiology-Paris.*

Friston, K. J. The free energy principle and counterfactuals: Linking perception, action and decision-making. (2023). *Frontiers in Psychology, 14,* 713955.

Friston, K.J. Harrison, L. and Penny, W., A theory of cortical responses. (2003). *Philosophical Transactions of the Royal Society B: Biological Sciences.*

Friston, K. J., Parr, T. and Stephan, K. E. Active inference and counterfactual thinking: A unification. (2022). *Neural Computation, 34*(1), 144-187.

Friston, K. J. and Stephan, K. E. Counterfactual inference and the free energy principle. (2022). *The Oxford Handbook of Cognitive Psychology* (pp. 451-474), Oxford University Press.

Friston, K. J. The free energy principle and counterfactuals: Linking perception, action, and decision-making. (2023). *Frontiers in Psychology, 14*, 713955.

Gibbons, G. W. and Shellard, E. P. The Temporal Topology of the Universe (https://www.google.com/search?q=2013).

Gurzadyan, V. and Penrose, R. Concentric circles in WMAP data may provide evidence of violent pre-Big-Bang activity (https://www.google.com/search?q=2010).

Penrose, R. Cycles of Time: An Extraordinary New View of the *Universe* (https://www.google.com/search?q=2010).

Penrose, R. Fashion, Faith, and Fantasy in the New Physics of the Universe (2016).

Kahneman, Daniel and Tversky, Amos. Heuristics and Biases: The Psychology of Intuitive Judgment (https://www.google.com/search?q=2002).

Chapter 2 - What Is Physical Reality?

1. Cat designed by freepick. Freepik.com
2. Griffiths, David J. (and Darrell F. Schroeter in later editions), Introduction to Quantum Mechanics (2018).
3. Gondran Alexandre, CC BY-SA 4.0 https://creativecommons.org/licenses/by-sa/4.0, via Wikimedia Commons
4. https://youtu.be/-1PsQIciMEc?si=FV8OBdCi43zG2TJx
5. https://www.youtube.com/watch?v=16kzFN0SWYg&t=1111s

Further readings

Linde, A.D. Particle Physics and Inflationary Cosmology. *Harwood Academic Publishers* (1990).

Chapter 3: Interpretations of Quantum Mechanics

1. Wigner E. The Scientist Speculates (I.J. Good, ed.).
2. Hossenfelder S. Existential Physics (2022).
3. Hossenfelder S. Lost in Math (2018).
4. https://www.youtube.com/watch?v=UlaaVp3F844

Chapter 7: Arrow of Time: Does Consciousness Steer it?

1. Gozdz, A. & Góźdź, Marek & Pędrak, Aleksandra. (Quantum Time and Quantum Evolution. (2023). *Universe*. 9. 256. 10.3390/universe9060256.

Chapter 8: Wignner's Friend: The Ultimate Observer Test

1. Lostaglio M., Bowles J. The original Wigner's friend paradox within a realist toy model: *The Royal Society.* https://doi.org/10.1098/rspa.2021.0273
2. Wang E. Quantum Bootcamp Part III: Eugene Wigner, His Friend(s), and Quantum: (2021) Princeton.
3. https://www.igogi-vienna.at/research/brukner-group/the-measurement-problem-and-wigners-friend-thought-experiment

Chapter 9: *SpooQ's!* Superposition Test

1. https://www.studysmarter.co.uk/explanations/physics/astrophysics/quantum-zeno-effect/
2. e93beb7d52ae#:~:text=Zeno's%20ancient%20arrow%20paradox%20has,the%20state%20of%20quantum%20systems

Chapter 12: *The Conscious-O-Meter*

1. https://geometrymatters.com/graph-theory-for-identifying-connectivity-patterns-in-human-brain-networks/
2. Tononi, G. An information integration theory of consciousness. (2004). BMC *Neuroscience, 5*(42).
3. Tononi, G. Consciousness as integrated information: a provisional manifesto. 2(008). *The Biological Bulletin, 215*(3), 216-242.
4. McCulloch, W. S. and Pitts, W. A logical calculus of the ideas immanent in nervous activity. (1943). B*ulletin of Mathematical Biophysics, 5*(4), 115-133.

5. Rosenblatt, F. The perceptron: A probabilistic model for information storage and organization in the brain. *Psychological Review, 65*(6), 386-40.

6. Hinton, G. E., Srivastava, N., Krizhevsky, A., Sutskever, I. and Salakhutdinov, R. R. Improving neural networks by preventing co-adaptation of feature detectors. (2012). *arXiv preprint arXiv:1207.0580.*

7. LeCun, Y., Bottou, L., Bengio, Y. and Haffner, P. Gradient-based learning applied to document recognition. (1998). *Proceedings of the IEEE, 86*(11), 2278-2324

8. Dayan, P., & Abbott, L. F. Theoretical neuroscience: computational and mathematical modeling of neural systems. (2001). *MIT press.*

9. Trappenberg, T. P. Fundamentals of computational neuroscience. *Oxford university press.*

10. O'Reilly, R. C., & Munakata, Y. Computational explorations in cognitive neuroscience: Understanding the mind by simulating the brain. 2000. *MIT press.*

11. https://www.dwavesys.com/resources

12. https://en.wikipedia.org/wiki/Adiabatic_theorem#:~:text=In%20simpler%20te rms%2C%20a%20quantum,but%20when%20subjected%20to%20rapidly&text=%2C%20remains%20in%20the%20ground%20state,to%20the%20slo wly%20varying%20conditions.

13. Albantakis, Larissa, Robert Prentner and Ian Durham. Computing the Integrated Information of a Quantum Mechanism. *Entropy 25*, no. 3: 449. (2023). https://doi.org/10.3390/e25030449

14. Hu X, Huang H, Peng B, Han J, Liu N, Lv J, Guo L, Guo C, Liu T. Latent source mining in FMRI via restricted Boltzmann machine: (2018). *Human Brain Mapp.*

Further reading

Siegel J. Quantum Mechanics and the puzzle of human consciousness. (2024). *Allen Institute.*

Tononi, G. Integrated information theory: An overview. (2012). *Progress in Brain Research.*

Tononi, G. The neural correlates of consciousness: An update on integrated information theory. (2012). *Progress in Brain Research.*

B. Baumard and F. Metzinger (Eds.), Conscious states: Computation, phenomenal experience and the binding problem. (2018). *Oxford University Press.*

Tononi, G. Integrated information theory: From consciousness to coma. (2012). Springer.

Tononi, G. and Koch, *C. Consciousness.*

Baars, B. J. A cognitive theory of consciousness. (1998). *Cambridge, MA: MIT Press.*

Graziano, M. S. A. Consciousness and the social brain. (2011). *Oxford University Press.*

IUCN Red List: https://www.iucnredlist.org/

Malakar, R. Information conservation theory of consciousness: A new perspective on the nature of reality. (2016). *Frontiers in Psychology, 7,* 1696.

Malakar, R. Information conservation theory of consciousness: A computational model. (2017). *Frontiers in Psychology, 8,* 1728.

Mancuso, S. and Viola, G. The conscious plant: A scientific exploration into the green kingdom. (2018). *MIT Press.*

Mora et al. study: https://www.pnas.org/doi/full/10.1073/pnas.162359199

Tononi, G. and Koch, C. Consciousness: Towards a science of the subjective. (https://www.google.com/search?q=2015). *Springer.*

Sejnowski TJ, Koch C, Churchland PS. Computational neuroscience. (1988). *Science.*

Rolls, Edmund T. On the Relation between the Mind and the Brain: A Neuroscience Perspective, *Philosophia Scientiae.*

Tyagi, Bhaumik. Understanding the Brain through Code and Mathematics: An Introduction to Computational Neuroscience (2023).

Nielsen, M. A., & Chuang, I. L. Quantum computation and quantum information. https://www.google.com/search?q=2010). *Cambridge University Press.*

Rieffel, E. G., & Polak, W. H. Quantum computing: A gentle introduction. (2011). *MIT Press.*

Epilogue

1. Arkani-Hamed, Nima. An interview with YouTube channel Closer to Truth.
2. Hossenfelder, Sabine. Existential Physics (2022).
3. Hossenfelder, Sabine. Lost in Math (2018).

www.ingramcontent.com/pod-product-compliance
Lightning Source LLC
Chambersburg PA
CBHW070938210326
41520CB00021B/6959